THE FUTURE OF INTELLIGENCE

Humans and Artificial Minds

Exploring the Relationship Between Human Thinking and Artificial Intelligence

By

Susie Hala

IntelliGloss AI Education Series

IntelliGloss Press

Legal Disclaimer

Educational Purpose Disclaimer

"This publication is part of the IntelliGloss AI Education Series, a structured curriculum support system for AI literacy in grades 6–12."

The information presented in this publication is provided for educational and informational purposes only. This book is intended to support educators, administrators, parents, and learners in understanding the concepts, opportunities, and challenges associated with artificial intelligence in education. It is not intended to serve as legal, technical, policy, cybersecurity, or professional advice.

Readers should consult qualified professionals, institutional guidelines, and local regulations before implementing any educational programs, technologies, or policies described in this book.

Technology and Accuracy Disclaimer

Artificial intelligence is a rapidly evolving field. While every effort has been made to ensure the accuracy and reliability of the information presented at the time of publication, technologies, policies, research findings, and best practices may change over time. The author and publisher make no guarantees that all information will remain current or applicable in future technological environments.

Readers are encouraged to verify tools, platforms, and recommendations independently before using them in educational settings.

Implementation Responsibility

Schools, educators, and institutions are solely responsible for determining how and whether to implement artificial intelligence tools, strategies, or classroom activities discussed in this book. Educational environments vary widely, and policies regarding technology use, student privacy, and curriculum development differ by district, state, and country.

The author and publisher assume no responsibility for outcomes resulting from the implementation of ideas, examples, or frameworks described in this book.

Student Safety and Data Privacy

Artificial intelligence tools may involve the collection or processing of data. Educators and institutions are responsible for ensuring that any technology used complies with applicable student privacy laws, data protection regulations, and school policies, including but not limited to FERPA, COPPA, GDPR, or other relevant regulations depending on jurisdiction.

The author and publisher are not responsible for the privacy practices, security policies, or data handling procedures of third-party platforms or software mentioned in this book.

Third-Party Tools and References
This book may reference third-party technologies, organizations, research studies, or platforms for illustrative and educational purposes. Such references do not constitute endorsements or guarantees regarding the performance, reliability, safety, or suitability of those tools for educational use.

Educators and institutions should conduct their own evaluations and risk assessments before adopting any technology.

No Professional Liability
The author and publisher shall not be held liable for any direct, indirect, incidental, or consequential damages resulting from the use or misuse of the information contained in this book. This includes but is not limited to decisions related to curriculum design, technology adoption, classroom instruction, institutional policies, or educational outcomes.

Educational Interpretation
The concepts and examples provided in this book are intended to support discussion and learning. They should not be interpreted as official standards, mandatory guidelines, or universally applicable policies for artificial intelligence education.

Schools and educators are encouraged to adapt the ideas presented here according to their local policies, cultural context, educational standards, and student needs.

Author's Intent
The goal of this book is to promote responsible AI literacy, thoughtful discussion, and ethical awareness as schools prepare students for a world increasingly shaped by intelligent technologies.

The author encourages readers to approach artificial intelligence education with curiosity, caution, and a commitment to responsible learning.

Title: THE FUTURE OF INTELLIGENCE Humans and Artificial Minds

Exploring the Relationship Between Human Thinking and Artificial Intelligence

ISBN: 978-1-972925-10-2

Published by
IntelliGloss Press
United States of America

Dedication

This book is dedicated to the next generation of learners—
the students who will grow up in a world shaped by intelligent technology,
and who deserve to understand it with clarity, confidence, and purpose.

To the educators who show up every day with dedication and heart,
guiding students through a rapidly changing world—
your role has never been more important.

To the parents and families who support learning at home,
encouraging curiosity, responsibility, and growth—
you are the foundation of every child's success.

And to my granddaughters,
whose curiosity, imagination, and future inspire me to do this work—
this book is for you.

May you not only learn how to use technology,
but understand it, question it, and shape it for the good of others.

About the Author

Susie Hala
Founder of the IntelliGloss AI Education Series

Susie Hala is an author, educator, and curriculum developer dedicated to advancing artificial intelligence literacy in education. Her work focuses on helping schools, educators, and students understand how intelligent technologies are shaping modern society—and why that understanding is essential for the next generation.

Hala is the creator of the IntelliGloss AI Education Series, a comprehensive and growing collection of educational books designed to make complex artificial intelligence concepts accessible, structured, and practical for classroom use. Her work emphasizes clarity, organization, and responsible implementation, enabling schools to introduce AI literacy without requiring advanced technical backgrounds.

In addition to her work in education, Hala brings hands-on experience working with artificial intelligence systems, including early exposure to rule-based AI technologies during her time working with AT&T. She has also developed modern digital platforms that integrate artificial intelligence, automation, and online systems, building e-commerce solutions and AI-driven applications that connect payment platforms, hosting environments, and conversational AI tools into real-world, functional systems.

This combination of foundational understanding and practical experience allows her to translate complex AI concepts into clear, accessible language for both educators and students.

Recognizing that artificial intelligence is rapidly transforming how information is created, interpreted, and used, Hala advocates for equipping students with the ability not only to use AI tools, but to understand, question, and evaluate them. Her work emphasizes critical thinking, ethical awareness, and responsible digital citizenship.

The IntelliGloss AI Education Series supports schools in teaching key areas such as artificial intelligence foundations, algorithms, machine learning concepts, and AI ethics, as well as the broader evolution of intelligent systems—from Artificial Narrow Intelligence (ANI) to Artificial General Intelligence (AGI) and Artificial Superintelligence (ASI). These resources are designed for students in grades 6–12 and for educators seeking structured, classroom-ready approaches to AI literacy.

In addition to her work in education publishing, Hala has a strong background in entrepreneurship, having built and managed multiple businesses and educational initiatives. Her experience across industries reinforces her belief that technology education must remain grounded in human values, critical thinking, and ethical responsibility.

Her mission is clear: to ensure that the next generation understands intelligent technology—not just how to use it, but how it truly works.

Preface

Two Students. Two Paths. One Future.

Before we begin, it is important to understand why learning how artificial intelligence works—not just how to use it—matters for every student.

Imagine two students who have just graduated from high school and are preparing for a job interview. The first student knows how to use artificial intelligence tools. This student can generate answers, complete assignments quickly, and rely on AI to assist with tasks. However, this student does not understand how artificial intelligence works behind the scenes.

The second student also knows how to use AI tools, but in addition, understands the foundational concepts behind them. This student has learned about data, algorithms, tokens, models, and how intelligent systems are trained and operate. This student can think critically about AI, recognize its limitations, and adapt when needed.

At first, the first student may appear faster and more efficient. However, over time, the second student demonstrates a deeper level of understanding, independence, and problem-solving ability. The second student is not only able to use artificial intelligence but also to question it, improve it, and grow with it.

This book is written for the second student.

In a world where artificial intelligence is becoming part of everyday life, students must not only learn how to use intelligent systems—they must learn how they work. True understanding leads to confidence, responsibility, and the ability to lead in an AI-driven future.

This perspective is not just theoretical—it is personal.

There was a time when I was the first student.

I knew how to use systems. I followed processes, completed tasks, and worked within structured environments that required accuracy and consistency. At the time, I did not think of these systems as artificial intelligence. They were simply part of the job—tools that had to be learned, followed, and mastered.

But looking back, I now understand something much deeper. I was working within rule-based systems—the earliest foundation of what we now recognize as artificial intelligence.

That realization changed everything.

As I began studying modern AI, including machine learning, deep learning, and intelligent systems, I started to see a clear connection between the past and the present. The structured systems I once worked in, the logic behind decision-making, and the step-by-step processes all still exist today. They have simply evolved.

Artificial intelligence did not appear overnight. It grew from rule-based systems into systems that can now learn, adapt, and assist in ways we never imagined.

I made a decision to move from simply using systems to understanding them.

That journey—from student one to student two—is the reason these books exist.

My mission is to help others bridge that same gap.

I believe that artificial intelligence should be understandable for everyone. It should not feel intimidating or out of reach. Students should feel confident exploring it. Teachers should feel prepared to introduce it. Everyday learners should have access to clear, structured explanations that make sense.

You do not need a background in technology. You do not need to know how to code. You only need the right explanation.

A Note from the Author

AI Is a Tool — Not a Decision Maker

Artificial Intelligence is becoming part of everyday life. Students use it to complete assignments, teachers use it to support instruction, and businesses use it to improve efficiency. While these tools are powerful and useful, there is an important truth that must remain clear from the very beginning:

Artificial Intelligence does not make decisions—people do.

One of the main purposes of this book is to help readers understand how AI works, what it can do, and—just as importantly—what it cannot do. AI systems are designed to process information, identify patterns, and generate responses based on data. They can assist with thinking, but they do not think independently. They can provide answers, but they do not understand consequences. They can suggest ideas, but they do not choose actions.

Every outcome that involves AI begins and ends with human involvement. A person asks the question. A person reviews the response. A person decides what to do next. Responsibility does not transfer to the machine.

In today's world, it is easy to confuse speed with intelligence and automation with authority. Because AI can respond quickly and confidently, it may appear as though it is making decisions. In reality, it is performing a task based on patterns it has learned—not exercising judgment.

This distinction matters. When people begin to rely on AI without understanding its role, they risk misunderstanding information, misusing technology, or placing trust in a system that is not capable of accountability. Clear understanding leads to responsible use.

This book was written to remove confusion and provide clarity. It is designed to help students and educators move beyond simply using AI tools and begin understanding the structure behind them. By learning how AI supports human thinking—rather than replacing it—readers can approach technology with confidence, awareness, and responsibility.

Artificial Intelligence is a powerful assistant. It can enhance learning, support creativity, and improve productivity. But it remains a tool.

The decisions, the actions, and the responsibility will always belong to the human.

— Susie Hala

Bringing Clarity to Artificial Intelligence

During a recent meeting with a school district, it became clear that one of the biggest challenges in artificial intelligence education is not the technology itself—but the way it is explained.

In that discussion, terms such as **chatbots, generative AI, agentic AI, autonomy, tools, hallucinations, and AI usage levels** were all mentioned together. While each term represents a distinct concept, they were often blended into one conversation without clear distinctions. The result was confusion, even among those who were actively trying to understand and make informed decisions.

This experience revealed an important truth:
Without clear definitions and structure, artificial intelligence becomes difficult to teach, difficult to learn, and difficult to use responsibly.

This book is designed to provide that structure.

Understanding the Core Concepts

To build a strong foundation, it is essential to separate and define the key ideas that are often confused.

Chatbot
A chatbot is a conversational interface that allows users to interact with an AI system through questions and prompts. It responds to input but does not act beyond what is directly requested.

Generative AI (Gen AI)
Generative AI refers to systems that create new content, including text, images, audio, and code. These systems generate outputs based on patterns learned from large datasets. While powerful, generative AI does not truly understand information and may occasionally produce inaccurate results.

Agentic AI
Agentic AI describes systems that can take action toward a goal. Rather than responding to a single prompt, these systems can plan, make decisions within defined limits, and carry out multi-step tasks. This introduces a higher level of functional independence.

Autonomy
Autonomy refers to how independently an AI system operates. Some systems require constant human guidance, while others can perform multiple steps with minimal input. Autonomy is not a separate type of AI, but a measure of how much control the system has in completing tasks.

Additional Terms That Shape Understanding

AI Tools
AI tools are the applications people use to interact with artificial intelligence, such as writing assistants, design platforms, and data analysis systems. These tools make AI accessible in everyday tasks.

Hallucination
A hallucination occurs when an AI system generates information that appears correct but is inaccurate or unsupported. This is a known limitation of generative systems and highlights the need for human verification.

The AI Use Continuum

Another important concept is how AI is used at different levels. Rather than viewing AI as a single capability, it is more accurate to understand it as a continuum of use:

Assistive Use
AI supports simple tasks such as answering questions, correcting grammar, or providing suggestions.

Augmented Use
AI helps expand human capability by generating ideas, organizing content, and assisting in problem-solving.

Agentic Use
AI begins to take on multi-step responsibilities, planning and completing tasks with reduced human input.

This continuum shows that AI can range from a helpful assistant to a more independent system, depending on how it is applied.

The Moment Everything Sounded Like One Conversation

During a recent discussion on artificial intelligence, I listened closely as different experts spoke about the future of AI. One referenced advanced systems beyond human intelligence, another discussed goal-driven AI systems, and the conversation shifted quickly between topics like autonomy, tools, and system behavior. At first, it sounded like a single, unified discussion. But the more I listened, the more I realized something important.

They were not all talking about the same thing.

One part of the conversation focused on future possibilities—high-level systems that do not yet exist. Another part focused on current technologies that are already being used today. At the same time, different terms were being used to describe how AI behaves, how it makes decisions, and how much independence it may have. These ideas were layered together in a way that made them sound connected, even when they were not.

This experience revealed a major challenge in understanding artificial intelligence today. Many discussions combine different types of AI, different levels of intelligence, and different behaviors into one conversation. Without clear structure, these ideas begin to blur together. For someone trying to learn, this can feel overwhelming and confusing.

This book is built to solve that problem.

Artificial intelligence is not one single concept. It is a system made up of multiple layers. Some layers describe how intelligent a system is. Other layers describe how the system behaves. Still others describe how much control or independence the system has. When these layers are separated and explained clearly, AI becomes much easier to understand.

The goal of this book is to bring clarity where confusion often exists. Instead of mixing everything into one explanation, each concept is presented in a structured and organized way. This allows students, educators, and readers to see how the pieces fit together without losing sight of what each part actually means.

Understanding artificial intelligence begins with understanding its structure.

Once that structure is clear, the conversation becomes clearer as well.

Why Clarity Matters

When these concepts are not clearly defined, they merge into what can be described as a "conceptual blur." This can lead to misunderstandings about what AI can and cannot do, as well as unrealistic expectations or misuse in educational settings.

Clear understanding helps ensure that:

Students remain active learners, not passive users

Educators maintain control over instructional outcomes

AI is used as a support tool rather than a replacement for thinking

The Purpose of This Book

This book is built on a simple but essential principle:

Students should not only use artificial intelligence—they should understand it.

By clearly defining core concepts such as chatbots, generative AI, agentic AI, autonomy, and the continuum of AI use, this curriculum provides a structured pathway for developing both technical understanding and critical thinking.

Artificial intelligence is not a single tool or system. It is a collection of capabilities that must be understood in context.

Clarity is the foundation of responsible innovation.

Structuring Artificial Intelligence to Create Clarity and Understanding

Why Structure Is Necessary

Artificial intelligence is often presented as one unified concept. In many discussions, different ideas such as intelligence level, system behavior, and autonomy are combined into a single explanation. This can make AI appear more complex and confusing than it actually is.

Clarity begins when these ideas are separated and organized.

Structuring artificial intelligence allows each concept to be understood on its own before being connected to the larger system.

The Problem Without Structure

When AI is not structured, conversations may include multiple terms that sound related but describe entirely different things. For example, a discussion may move quickly between advanced future systems, current tools, and system behaviors without clearly distinguishing between them.

This creates the impression that all forms of AI operate at the same level and in the same way, which is not accurate.

Without structure, understanding becomes difficult.

The Solution: A Structured Approach

Artificial intelligence becomes clearer when it is organized into distinct categories.

Level explains how intelligent a system is.
Behavior explains how the system operates.
Control explains how much independence the system has.

Each category answers a different question. Together, they form a complete picture of how AI systems function.

Clarity Through Separation

A system can operate at a current level of intelligence while demonstrating different behaviors or levels of independence. For example, a system that generates responses and a system that takes steps toward completing a task may operate at the same intelligence level but differ in how they behave.

This distinction becomes clear only when structure is applied.

Instructional Insight

When teaching artificial intelligence, it is important to avoid presenting all concepts at once. Separating ideas into categories allows learners to focus on one dimension at a time, building understanding step by step.

Structured learning leads to deeper comprehension.

Key Takeaway

Artificial intelligence is not one single idea. It is a system made up of multiple parts. Structuring those parts creates clarity, improves understanding, and allows learners to see how the system truly works.

Reflection Question

How does separating intelligence level, behavior, and control make artificial intelligence easier to understand?

The IntelliGloss Instructional Approach

The IntelliGloss AI Education Series is designed to help students move beyond simply using AI tools and begin understanding artificial intelligence as a complete system.

Students learn how AI is built, how it operates, and how it connects to the real world.

A Message to Students and Readers

If you have ever felt that technology was too complicated…
If you have ever been confused by systems that were never clearly explained…

I want you to know:

You are capable of understanding artificial intelligence.

Sometimes, the challenge is not the concept itself—it is how it has been presented.

This series was created to change that.

A Message to Educators

To the educators using this material:

These books are designed to support you in introducing artificial intelligence in a way that is clear, structured, and meaningful for students.

AI is becoming part of every industry, and students deserve more than just exposure to tools—they deserve understanding.

This series aims to provide that foundation.

Final Reflection

Every experience we have—especially the challenging ones—prepares us for something greater.

The systems I once worked in taught me discipline, structure, and attention to detail. Today, those same lessons allow me to break down complex ideas and make them understandable for others.

That is what this series represents:

Turning complexity into clarity.

Why Artificial Intelligence Education Matters for Every Student

Why AI Literacy Is Essential for Every Student

Artificial intelligence is no longer a future concept or a specialized technology used only by experts. It is already part of everyday life for students of all ages. From smartphones and search engines to navigation apps, social media, streaming platforms, and educational tools, AI systems influence how information is accessed, decisions are made, and learning takes place.

Because AI is embedded in daily life, understanding it is no longer optional. It is a foundational skill, similar to digital literacy and critical thinking. Students do not need to become programmers or engineers, but they do need to understand what AI is, how it works at a basic level, and how it affects their lives.

AI education should not be limited to computer science classrooms. Every student interacts with AI systems daily. Teaching students how these systems function empowers them to become informed users rather than passive consumers of technology.

This book is designed to make artificial intelligence understandable, approachable, and relevant. It focuses on clarity, responsibility, and real-world understanding to help students prepare for a future where humans and intelligent systems work side by side.

IntelliGloss AI Education Series

Teaching how AI actually works—from hardware to power to global systems.

Why AI Education Matters

Artificial intelligence is already shaping how students learn, write, research, and solve problems.

Yet many students—and educators—are using AI tools without understanding how these systems actually work.

Schools need more than access to technology.
They need structured, responsible AI education.

Our Approach

The IntelliGloss AI Education Series is designed to help students move beyond simply using AI tools and begin understanding artificial intelligence as a complete system.

Students learn how AI is built, how it operates, and how it connects to the real world.

The IntelliGloss Framework

Our books introduce AI through a clear and connected system:

Chip Family — the hardware that powers AI

Watts Family — the rate at which AI uses power

Energy Family — the total power AI consumes over time

Infrastructure Family — the global systems that support AI

This framework helps students understand AI from the inside out.

Built for the Classroom

Each book includes:

Structured chapters aligned with semester learning

Clear explanations designed for student understanding

Reflection sections with learning objectives

Worksheets, quizzes, and answer keys

Teacher Implementation Guides

Supporting Schools and Educators

The IntelliGloss series supports:

Responsible and ethical AI use

Student data awareness and safety

Critical thinking and digital literacy

Classroom-ready instruction for grades 6–12 and beyond

What This Book Is—and What It Is Not

Many people today are becoming familiar with artificial intelligence through tools like ChatGPT, Microsoft Copilot, Canva, and other AI-powered applications. These tools can help with writing, answering questions, creating images, and completing everyday tasks.

In addition to these tools, you may also hear terms such as **AI agents, agentic AI, autonomous systems, and generative AI**. While these may sound complex, they are all part of the same evolving landscape of artificial intelligence technologies.

This book is not a guide on how to use those tools.

Instead, it explains what makes those tools possible.

Behind every AI tool is a system made up of multiple layers working together. At the most basic level, information is represented as bits and processed through powerful computer chips. That information is organized into data, broken into tokens, and then processed by algorithms and models that allow the system to recognize patterns, generate responses, and perform tasks.

These components—chips, bits, tokens, data, algorithms, and models—work together as a complete system. While users may only see the tool, the real intelligence comes from how these underlying parts interact behind the scenes.

Artificial intelligence can also be understood in terms of different levels of capability. Most of the AI systems used today fall under **Artificial Narrow Intelligence (ANI)**, which is designed to perform specific tasks, such as answering questions or generating content. More advanced forms, such as **Artificial General Intelligence (AGI)** and **Artificial Superintelligence (ASI)** represent future possibilities where AI could match or exceed human-level thinking. While these advanced levels are still theoretical, understanding the differences helps provide a broader view of where AI is today and where it may be heading.

For students, parents, and educators, this distinction is important. Learning how to use AI tools is helpful, but understanding how they work—and where they fit within the larger AI landscape—builds stronger thinking, better decision-making, and more responsible use.

Most importantly, artificial intelligence does not make decisions on its own. It does not take responsibility for actions or outcomes. AI systems generate responses based on data and patterns, but it is always the human who decides what to do with that information.

This book is designed to make these underlying systems clear and accessible, even for readers who are new to artificial intelligence.

Artificial intelligence is still in its early stages of development. As these technologies continue to grow, those who understand the foundation behind them will be better prepared to adapt, question, and use them wisely.

Who This Book Is For

This book is designed for educators, school leaders, and curriculum developers guiding students in an age of rapidly advancing technology.

Artificial intelligence is already influencing how students learn, research, and communicate. Yet many educators have not received formal training in how these systems work or how they should be addressed in the classroom.

This book helps bridge that gap by providing clear explanations and practical frameworks for understanding artificial intelligence in an educational setting.

How Schools Can Use This Book

This book supports schools in developing a structured approach to artificial intelligence literacy.

It can be used for:

- Professional development for educators
- Classroom discussions and activities
- Curriculum support across subject areas
- Policy and academic integrity conversations

Rather than focusing on specific tools, this book emphasizes lasting principles that help students understand, evaluate, and use AI responsibly.

Closing Perspective

Artificial intelligence will continue to evolve. With the right knowledge and leadership, schools can ensure that students not only use these technologies, but understand them, question them, and apply them responsibly.

IntelliGloss is more than a book series. It is a structured approach to preparing students for a future shaped by artificial intelligence.

Rather than focusing only on how to use AI tools, this series helps students understand how AI actually works—from its foundational building blocks to real-world applications.

“Before exploring the technical structure of artificial intelligence, it is equally important to understand the human principles that guide its use.”

Human Values Family

Definition

The Human Values Family in artificial intelligence represents the guiding principles that shape how AI systems are created, used, and integrated into society. These values include ethics, responsibility, trust, and human well-being.

Artificial intelligence may be built on data, algorithms, and computational systems, but its true impact is measured by how it affects people. For this reason, AI is not only a technical system—it is a human-centered system.

Key Concepts

The Human Values Family exists to ensure that artificial intelligence serves humanity in a positive and responsible way. While AI systems can process information and generate outcomes, they do not possess judgment, empathy, or moral understanding.

This means that human values must guide every stage of AI development—from design to deployment.

AI reflects the intentions, decisions, and assumptions of the people who build and use it. If those values are not carefully considered, AI systems can unintentionally produce biased, harmful, or misleading results.

Understanding human values in AI is not optional—it is essential.

Core Elements of the Human Values Family

Ethics
Ethics in AI refers to making decisions that are fair, just, and respectful of human rights. This includes minimizing bias, protecting privacy, and ensuring transparency in how systems operate.

Responsibility
Responsibility means that humans remain accountable for AI systems. Developers, educators, organizations, and users must take ownership of how AI is used and the outcomes it produces.

Trust
Trust is critical for the successful adoption of AI. People must feel confident that AI systems are reliable, safe, and aligned with their best interests.

Human Well-Being
AI should enhance human life—not replace, harm, or diminish it. This includes supporting learning, improving access to information, and creating opportunities while maintaining human dignity.

Real-World Connection

Artificial intelligence is already influencing how people learn, communicate, and make decisions. From search engines to classroom tools, AI is shaping daily experiences.

Without clear human values guiding these systems, there is a risk of reinforcing misinformation, bias, or unrealistic expectations.

For students, this means learning not only how to use AI tools, but how to question them, evaluate their outputs, and understand their limitations.

Applications in Education

In the classroom, the Human Values Family helps students develop critical thinking and ethical awareness. Teachers can use this framework to guide discussions about fairness, responsibility, and the impact of technology on society.

Students can explore questions such as:

How should AI be used responsibly?

What makes an AI system fair?

How does technology influence human behavior?

This approach supports both technical understanding and responsible decision-making.

Benefits

Teaching human values in AI helps students:

Develop ethical awareness

Strengthen critical thinking

Understand the societal impact of technology

Become responsible users and future creators of AI

It ensures that learning about AI goes beyond functionality and includes responsibility.

Challenges

Human values are not always universal. Different cultures, communities, and individuals may have different perspectives on what is considered fair or ethical.

Additionally, AI technology is evolving rapidly, often faster than policies and guidelines can keep up. This makes it essential for educators and students to stay informed and adaptable.

Summary

The Human Values Family reminds us that artificial intelligence is not just about machines—it is about people.

By focusing on ethics, responsibility, trust, and human well-being, students can learn to use AI thoughtfully and responsibly.

Understanding these principles helps prepare learners to navigate a world where artificial intelligence plays an increasingly important role in everyday life.

Executive Summary

IntelliGloss AI Education Series

Teaching How Artificial Intelligence Actually Works

Artificial intelligence is already shaping how students learn, write, research, and make decisions. Yet in many classrooms, students are being introduced to AI tools without understanding how these systems actually function.

This creates a critical gap.

Students may become efficient users of technology, but without foundational knowledge, they are not equipped to think critically, evaluate outputs, or use AI responsibly. As AI continues to expand across every industry, this gap will only grow more significant.

The **IntelliGloss AI Education Series** was created to address this challenge.

The Problem

Most current approaches to AI in education focus on tool usage rather than system understanding. As a result:

- Students rely on AI without understanding its limitations
- Educators lack structured frameworks for teaching AI concepts
- Key terms such as generative AI, agentic AI, and autonomy are often introduced without clear definitions
- Responsibility and ethical use are not consistently emphasized

Without clarity, artificial intelligence becomes difficult to teach, difficult to learn, and difficult to manage in a school environment.

The Solution

The IntelliGloss AI Education Series provides a structured, classroom-ready approach to artificial intelligence literacy for grades 6–12.

Rather than focusing on individual tools, this series teaches students how AI works as a complete system—from foundational components to real-world applications.

Students learn to move beyond surface-level interaction and develop true understanding.

The IntelliGloss Framework

The series introduces AI through a clear and connected structure, including:

Foundational building blocks such as data, algorithms, tokens, and models

System-level understanding through categories such as intelligence level, behavior, and control

Real-world context through applications, limitations, and responsible use

Human-centered learning through ethics, responsibility, trust, and student safety

This structured approach transforms artificial intelligence from a confusing concept into an understandable system.

Instructional Design

Each book in the series is designed for practical classroom use and includes:

Structured chapters aligned with instructional pacing

Clear, student-friendly explanations

Reflection sections to reinforce understanding

Worksheets, quizzes, and answer keys

Teacher implementation support

No prior technical background is required for educators or students.

Guiding Principle

A central message of the IntelliGloss series is:

Artificial Intelligence is a tool—not a decision maker.

Students learn that while AI can support thinking and provide information, all decisions, actions, and responsibilities remain with humans. This principle reinforces critical thinking, accountability, and responsible technology use.

Why It Matters

AI literacy is no longer optional. It is a foundational skill.

Students who understand how AI works will be better prepared to:

Think critically about information

Adapt to emerging technologies

Use AI responsibly and ethically

Succeed in an increasingly AI-driven world

Our Goal

The IntelliGloss AI Education Series is designed to help schools move beyond simply using artificial intelligence tools and toward a deeper understanding of how these systems actually work.

By providing clear structure, defined concepts, and real-world context, this series equips students and educators with the knowledge needed to think critically, evaluate information, and use AI responsibly.

The goal is not just to introduce technology—but to build understanding, confidence, and accountability in an AI-driven world.

This is not just about learning to use AI—it is about preparing students to understand it, question it, and lead with it.

Clarity leads to confidence.
Understanding leads to responsible use.

Mission Statement

The mission of the IntelliGloss AI Education Series is to make artificial intelligence clear, structured, and accessible for every student and educator.

In a world where AI is rapidly transforming how information is created, used, and understood, this series is designed to move beyond surface-level interaction and provide a deeper understanding of how intelligent systems actually work.

Through clear explanations, organized frameworks, and real-world context, IntelliGloss equips learners with the ability to think critically, evaluate information, and engage with technology responsibly.

This mission is grounded in a simple belief:
students should not only learn how to use artificial intelligence—they should understand it.

By building both technical awareness and human-centered understanding, the IntelliGloss AI Education Series prepares students to navigate, question, and shape a future increasingly influenced by intelligent technologies.

The goal is not just to keep up with artificial intelligence—but to understand it, guide it, and use it responsibly.

Vision Statement

The vision of the IntelliGloss AI Education Series is to create a future where artificial intelligence is clearly understood, responsibly used, and thoughtfully integrated into education and everyday life.

In this future, students are not passive users of technology, but informed thinkers who understand how intelligent systems work, recognize their limitations, and apply them with purpose and responsibility.

Educators are equipped with clear frameworks and structured resources that make artificial intelligence accessible, teachable, and meaningful across all subjects—not just in technical fields.

Schools become environments where technology is not simply adopted, but understood—where students are encouraged to question, evaluate, and engage with AI in ways that support learning, creativity, and ethical awareness.

As artificial intelligence continues to evolve, this vision supports a generation that is prepared not only to adapt to change, but to lead it with knowledge, confidence, and integrity.

The future of artificial intelligence will not be defined by technology alone—but by the understanding and responsibility of those who use it.

Connecting AI Understanding to Cybersecurity Awareness

Definition

Understanding artificial intelligence is the first step toward understanding cybersecurity. While cybersecurity focuses on protecting systems, data, and users, artificial intelligence explains how those systems are built, how they function, and where they may be vulnerable.

The Connection

Every AI system is made up of components such as data, algorithms, models, and infrastructure. These same components are also the targets of cybersecurity threats.

When students understand:

How data is collected and stored

How algorithms make decisions

How AI models process information

How systems operate across frontend and backend layers

They begin to recognize where risks can exist and how systems can be misused.

Why This Matters

In today's digital world, students are not just users of technology—they are participants in complex systems powered by artificial intelligence.

Without understanding how these systems work:

Technology becomes something they trust without question

Risks become harder to recognize

Decisions are made without awareness of consequences

With understanding:

Students become more aware of how their data is used

They recognize how systems can be manipulated

They make more informed and responsible choices

Building Digital Readiness

Digital readiness is not only about using tools. It is about understanding the systems behind those tools.

By learning how AI works, students develop:

Awareness of digital environments

Confidence in navigating technology

A foundation for future learning, including cybersecurity

Conclusion

Cybersecurity begins with understanding.

Before students can protect systems, they must first understand how those systems are built and how they operate. This is why learning artificial intelligence is not only about innovation—it is also about responsibility.

The Foundation of Artificial Intelligence

Definition

The foundation of Artificial Intelligence (AI) refers to the core building blocks that allow AI systems to function. These building blocks work together to help machines learn from information, recognize patterns, and make decisions. AI is not magic—it is a system built on structured components that operate behind the scenes.

Understanding the Foundation in Simple Terms

Artificial Intelligence can be compared to a house. Before a house is complete, it must be built on a strong foundation. This foundation includes concrete, wiring, plumbing, and support structures that are not always visible but are essential for the house to function properly.

In the same way, AI systems rely on foundational components that are not visible to users but are necessary for AI tools to work. When people use AI applications such as chatbots or design tools, they are interacting with the surface. The true power of AI comes from what is underneath.

Key Components of the AI Foundation

1. Data (The Information)

Data is the starting point of all AI systems. It includes text, images, numbers, audio, and other forms of information. AI systems learn by analyzing patterns within this data.

2. Algorithms (The Instructions)

Algorithms are step-by-step instructions that tell the AI system how to process data. They guide how the system learns, identifies patterns, and produces results.

3. Models (The Trained System)

A model is created when data and algorithms are combined through a training process. The model acts as the "brain" of the AI system, allowing it to make predictions and generate responses.

4. Computing Power (The Engine)

AI systems require powerful computers to process large amounts of data. This includes specialized hardware such as processors and graphics processing units (GPUs) that allow AI to function efficiently.

5. Infrastructure (The Environment)

Infrastructure refers to the systems that support AI operations, including servers, storage, networks, and energy systems. Data centers are a key part of this infrastructure, providing the physical space where AI systems operate.

How These Components Work Together

AI systems function by combining all foundational components into a single process. Data is collected and analyzed using algorithms. This process creates a model that can recognize patterns and make decisions. The entire system is powered by computing machines and supported by infrastructure.

Why Understanding the Foundation Matters

Many people use AI tools without understanding how they work. This can lead to confusion, misuse, or over-reliance on technology. By understanding the foundation of AI, students and educators can:

Develop critical thinking skills

Use AI tools more responsibly

Recognize the limitations of AI systems

Build a deeper understanding of modern technology

Real-World Example

When a student asks a chatbot a question, the response does not come from "thinking" like a human. Instead, the AI system uses a trained model that has learned patterns from large amounts of data. It follows algorithms to predict the most appropriate answer, using powerful computers and infrastructure to deliver the response quickly.

This process is also influenced by how the question is asked. The words, structure, and clarity of a question—often referred to as a *prompt*—can affect the quality of the response. Learning how to communicate effectively with AI systems is an important skill known as prompt engineering. For further understanding, students and educators may explore *What to Say When You Talk to Yourself: A Guide to Prompt Engineering* by Susie Hala, which explains how carefully structured prompts can guide AI responses more effectively.

Conclusion

The foundation of Artificial Intelligence is built on data, algorithms, models, computing power, and infrastructure. These components work together to create systems that can learn, predict, and assist humans. Understanding this foundation helps students move beyond simply using AI tools and toward truly understanding how artificial intelligence works.

The Foundational Families of Artificial Intelligence

Introduction

Artificial intelligence is often described as a powerful technology, but in reality, it is not a single system. It is a combination of multiple components working together to create intelligent behavior.

To better understand how artificial intelligence works, it is helpful to break it down into organized groups. In the IntelliGloss framework, these groups are called **Foundational Families of Artificial Intelligence**.

Each family represents a key part of how AI systems are built and operate. Some families focus on how information is created and stored. Others focus on how machines learn, make decisions, process language, or perform tasks. Additional families represent the physical hardware, energy systems, and global infrastructure that make artificial intelligence possible.

When these families work together, they form a complete AI system.

This approach allows students and educators to move beyond simply using artificial intelligence tools and begin to understand what is happening behind the scenes. Instead of viewing AI as a mystery, learners can see it as a structured system made up of clearly defined components.

The diagram on the next page provides a visual overview of these families and how they connect to form the foundation of artificial intelligence.

By learning each family step by step, students develop a deeper understanding of how AI systems process information, learn from data, and interact with the world.

Understanding the Foundational Families of Artificial Intelligence is the first step toward true AI literacy.

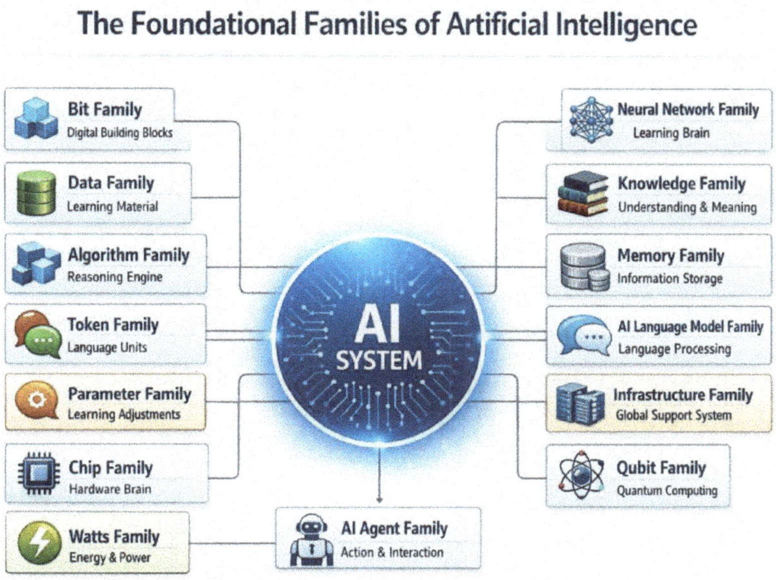

The Foundational Families of Artificial Intelligence

Family	Role	Description
Bit Family	Digital Building Blocks	The smallest unit of digital information (0 or 1). All computing systems begin with bits.
Algorithm Family	Reasoning Engine	Step-by-step instructions that tell computers how to solve problems and process information.
Token Family	Language Units	Pieces of text used by AI models to understand and generate human language.
Parameter Family	Learning Adjustment System	Adjustable numerical values inside AI models that allow them to learn patterns from data.
Data Family	Learning Material	The information AI studies in order to learn patterns, relationships, and knowledge.

Family	Role	Description
Logic Family	Decision Rules	The reasoning structures that allow AI systems to make decisions and evaluate conditions.
Memory Family	Information Storage	Systems that store and retrieve information so AI models can access past knowledge and context.
Knowledge Family	Understanding & Meaning	Structured information that allows AI to represent facts, relationships, and concepts.
Neural Network Family	Learning Brain	Interconnected layers of artificial neurons that enable AI systems to recognize patterns, learn from data, and make predictions.
AI Model Family	Intelligence Engine	The trained system that uses data, parameters, and neural networks to perform tasks such as prediction, classification, and generation.
AI Translator Architecture Family	Language Conversion System	Systems that transform input into output, enabling AI to convert text, speech, or images into meaningful responses.
AI Agent Family	Action and Task Execution System	AI systems designed to perceive information, make decisions, and carry out tasks autonomously or semi-autonomously.
Chip Family	Hardware Brain	Physical processors (GPUs, TPUs, CPUs, NPUs) that perform AI computations.
Watts Family	Energy Power	The electrical power required to run AI hardware and data centers.
AI Energy Family	Energy Source System	The total energy consumed over time to operate AI systems, including training and real-time usage.
Infrastructure Family	Global Support System	Data centers, internet networks, cloud systems, and global infrastructure that allow AI to operate and scale.
Ingredient Family	System Composition	The essential components that come together to create an AI system, including data, algorithms, models, hardware, and energy.
Qubit Family	Quantum Information Units	Quantum bits that can represent multiple states simultaneously, forming the foundation of quantum computing.

The Bit Family

Introduction

The Bit Family represents the starting point of all digital technology. Every computer system, application, and artificial intelligence model is built from bits—the smallest unit of information in computing.

A bit can hold one of two values: 0 or 1. While this may seem simple, these two values form the foundation of everything digital. When bits are combined and organized, they create larger units of data that allow computers to store text, display images, play videos, and run complex systems.

From a single bit to massive data systems, all digital information follows this same structure. This is how computers are able to represent and process the world in a form they can understand.

In artificial intelligence, bits are essential because they store the data that models learn from, process the calculations that drive decision-making, and support the systems that generate responses.

Understanding the Bit Family helps students see how simple binary signals grow into powerful technologies. It reveals that behind every advanced AI system is a foundation built from the most basic building blocks of information.

Bit Family — Visual Understanding of Data Size

How Much Data Can Each Unit Hold? (Simple Examples)

Unit	Size	Photos (Approx.)	Video (Approx. Hours)
Bit	1 bit	N/A	N/A
Byte	8 bits	1 character	N/A
Kilobyte (KB)	~1,000 bytes	Part of a paragraph	N/A
Megabyte (MB)	~1,000 KB	~1 photo	~1 minute of video
Gigabyte (GB)	~1,000 MB	~250 photos	~1–2 hours of video
Terabyte (TB)	~1,000 GB	~250,000 photos	~1,000–2,000 hours of video
Petabyte (PB)	~1,000 TB	~250 million photos	~1–2 million hours of video
Exabyte (EB)	~1,000 PB	~250 billion photos	~1–2 billion hours of video
Zettabyte (ZB)	~1,000 EB	~250 trillion photos	~1–2 trillion hours of video
Yottabyte (YB)	~1,000 ZB	~250 quadrillion photos	~1–2 quadrillion hours of video

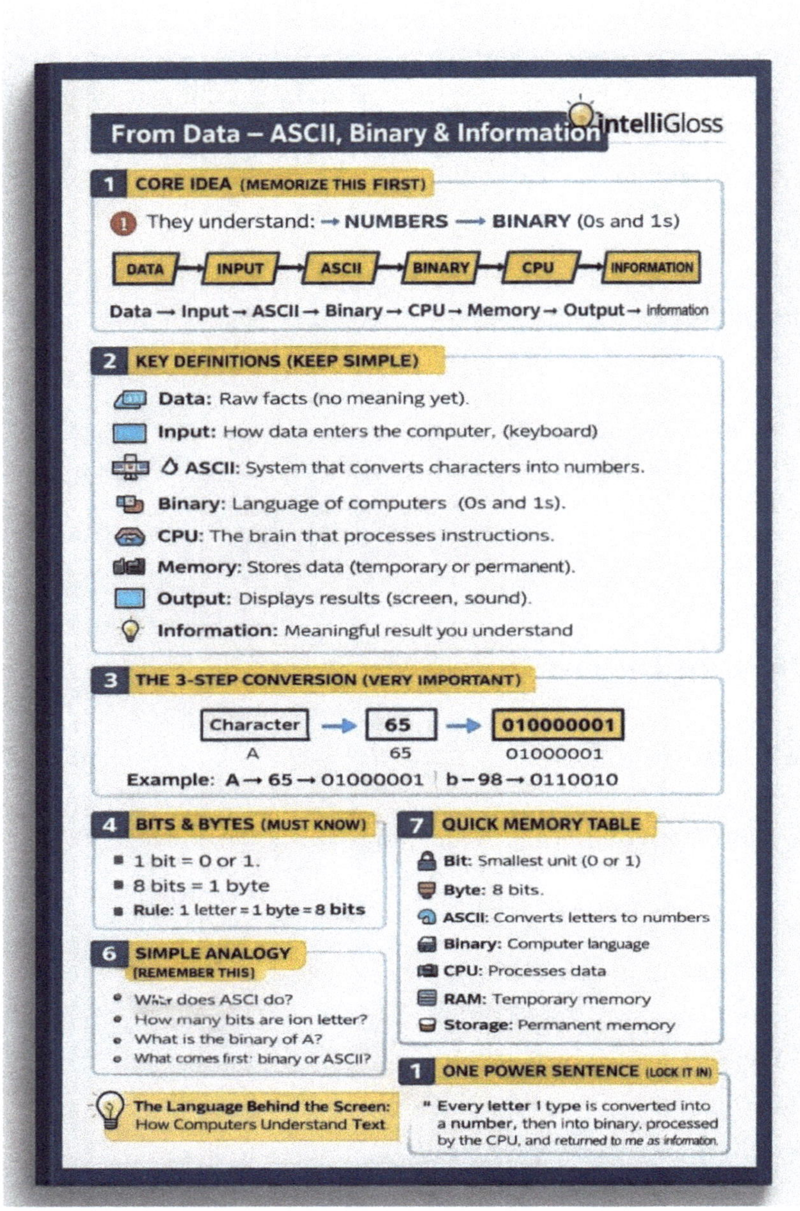

The Algorithm Family

Definition and Overview

The **Algorithm Family** represents the *thinking engine* of artificial intelligence — the set of mathematical instructions that guide how AI analyzes data, learns patterns, and makes decisions.

An **algorithm** is a sequence of steps designed to solve a problem or complete a task. In AI, algorithms tell the system *how to learn*, *how to predict*, and *how to improve* with every iteration.

In simple terms:

The Algorithm Family is the *logic of intelligence* — the invisible code that transforms data into decisions and predictions into progress.

How Algorithms Power AI

Every AI system — from simple chatbots to large-scale deep learning models — relies on algorithms as its foundation.
They determine how the model:

Processes and cleans data,

Identifies relationships between variables,

Learns from feedback, and

Improves over time.

Without algorithms, even the most powerful dataset or neural network would remain static and lifeless.

In essence, **algorithms are the teachers** that train AI how to think.

Why the Algorithm Family Matters

Algorithms are what separate automation from intelligence.
They allow machines to **reason**, **adapt**, and **learn** — not just follow commands.
For prompt engineers, understanding algorithmic behavior helps explain *why* AI sometimes misunderstands a question, repeats patterns, or changes tone after feedback.

Every AI response you receive follows an underlying process — a pattern recognition journey from input to output.

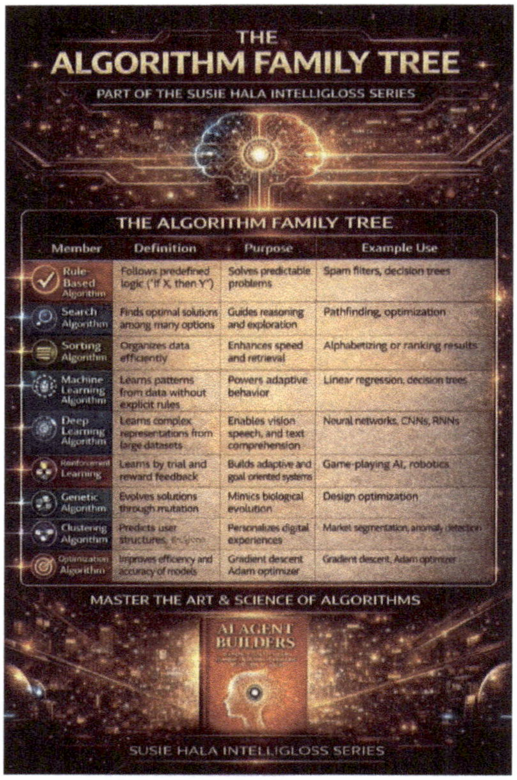

Analogy for Everyday Readers

Imagine AI as a chef in a kitchen:

The **Data Family** provides the ingredients.

The **Knowledge Family** gives the recipes.

> The **Algorithm Family** is the *cooking process* — mixing, heating, tasting, and adjusting until the dish is perfect.

Just as cooking methods determine the quality of food, algorithms determine the quality of intelligence.

Algorithm Efficiency and Energy Use

Not all algorithms are equal.
Some are **fast but shallow**, while others are **deep but energy-intensive**.
Training large AI models requires algorithms that can handle billions of calculations per second — powered by the **Watts Family** (energy) and executed through the **Chip Family** (hardware).

The design of efficient algorithms directly affects sustainability in AI — minimizing power consumption while maximizing accuracy and speed.

Algorithms and Prompt Engineering

For prompt engineers, algorithms explain *how* the AI arrives at its answers.
When a model gives inconsistent or off-topic results, it's often due to:

Misinterpretation of input signals (input encoding),

Bias in training algorithms, or

Overfitting to certain data patterns.

Knowing this helps you craft prompts that guide the AI's decision-making — not just its output.

💬 Quick Takeaway: The Algorithm Family is the reasoning heart of AI.
It transforms information into understanding and enables machines to think, learn, and evolve like digital problem-solvers.

Token Family Introduction

The Token Family represents one of the most important bridges between human language and artificial intelligence. While humans communicate using words, sentences, and meaning, AI systems do not understand language in the same way. Instead, they rely on tokens—small units of text that allow machines to process, analyze, and generate language step by step.

A token can be a word, part of a word, a character, or even punctuation. When a student types a sentence into an AI system, that sentence is not read as a whole idea. It is first broken down into tokens. These tokens become the input that the AI model uses to recognize patterns, predict outcomes, and generate responses.

The Token Family works closely with several other foundational families in artificial intelligence. It connects directly with the Data Family, which provides the text used for training, and the Neural Network Family, which processes token patterns to produce intelligent outputs. It also plays a key role in the AI Translator Architecture Family, where tokens are transformed, analyzed, and reassembled into meaningful responses.

Understanding tokens helps students see what is happening behind the scenes when they interact with AI tools. Instead of viewing AI as a "magic system," they begin to understand that every response is built from sequences of tokens processed through mathematical models.

In simple terms, if language is what humans speak, tokens are what AI understands.

Token Family Breakdown Chart

Component	Role	Description	Example
Tokenization	Input Conversion	The process of breaking text into smaller units (tokens) that AI can process	"Artificial Intelligence" → "Artificial" + "Intelligence"
Tokens	Language Units	The individual pieces of text used by AI models for understanding and generation	"AI", "learn", "ing", "."
Subword Tokens	Efficiency Units	Words split into smaller parts to handle unknown or complex vocabulary	"unbelievable" → "un", "believ", "able"
Character Tokens	Fine-Grained Units	Text broken down into individual characters for detailed processing	"AI" → "A" + "I"
Vocabulary (Token Set)	Token Library	The complete set of tokens an AI model recognizes and uses	GPT models have thousands to millions of tokens
Encoding	Token Mapping	Converting tokens into numerical representations for computation	"AI" → [1234, 5678] (example IDs)
Decoding	Output Reconstruction	Converting numerical outputs back into readable text	[1234, 5678] → "AI system"
Context Window	Memory Limit	The number of tokens an AI can process at one time	Example: 8,000 tokens ≈ several pages of text
Token Embeddings	Meaning Representation	Mathematical vectors that capture relationships between tokens	"king" and "queen" have similar embeddings
Token Prediction	Language Generation	The process of predicting the next token in a sequence	"AI is" → predicts "powerful"

The Token Family reminds us that behind every intelligent response is a structured sequence of tokens working together. What feels like natural conversation to humans is, for AI, a carefully processed stream of language units transformed into meaning.

The AI Ecosystem Family

The AI Ecosystem Family is a connected group of artificial intelligence technologies, systems, behaviors, and applications that work together to help machines perform intelligent tasks. This ecosystem includes core AI systems, AI agents, agentic AI, AI autonomy, autonomous AI workers, and generative AI. Each part plays a different role in how modern AI systems think, learn, create, make decisions, solve problems, and interact with the world.

At the center of the ecosystem is artificial intelligence itself — the broad field focused on building machines and software that can perform tasks that normally require human intelligence. AI agents act as the systems that carry out goals and actions. Agentic AI describes the ability of AI systems to plan, reason, adapt, and pursue objectives with limited human guidance. AI autonomy represents the level of independence an AI system has while making decisions and taking actions. AI autonomous workers apply these capabilities to perform real-world tasks like digital employees or assistants. Generative AI adds the ability to create new content such as text, images, music, video, code, and designs.

The AI Ecosystem Family demonstrates that modern artificial intelligence is no longer just one single technology. Instead, it is an interconnected environment of systems, tools, intelligence, automation, creativity, and human collaboration. As AI continues to evolve, these technologies are becoming increasingly integrated into education, business, healthcare, communication, cybersecurity, research, and everyday life.

Understanding the AI Ecosystem Family helps students, educators, and readers see how the different parts of AI connect together to shape the future of intelligent systems and the digital world.

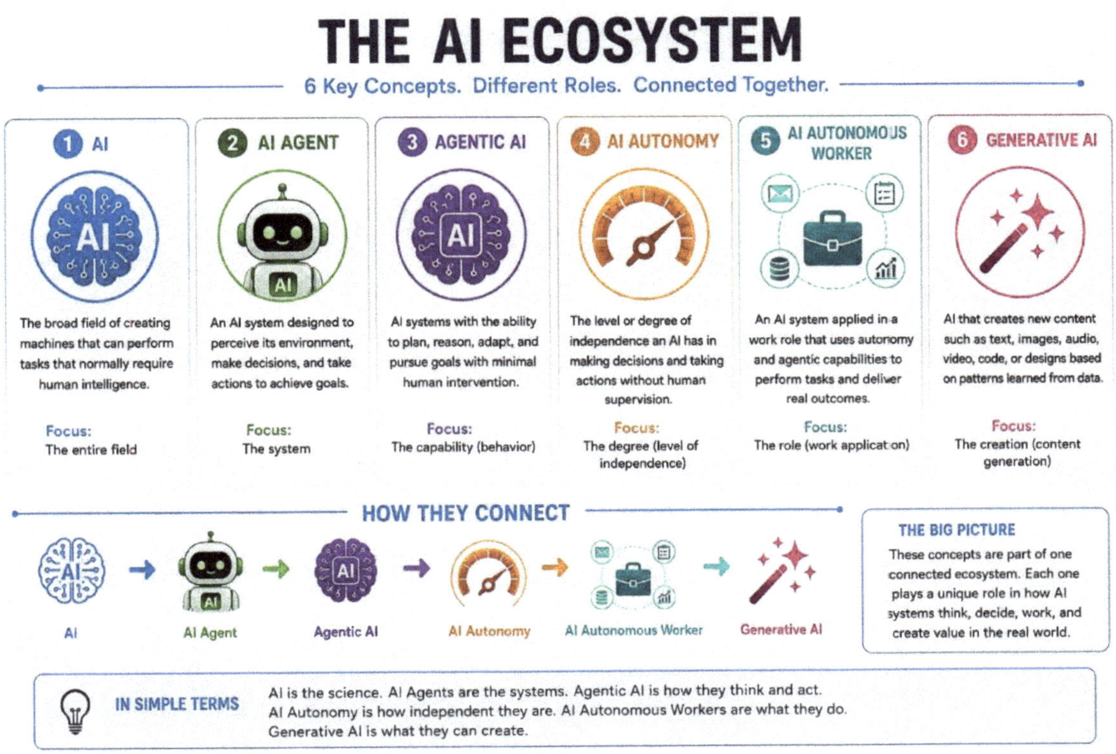

Parameter Family Introduction

The Parameter Family represents the internal learning system of artificial intelligence. While tokens provide the language and data provides the learning material, parameters are what allow AI systems to adjust, improve, and make intelligent decisions over time.

In simple terms, parameters are numerical values inside an AI model that change as the system learns. These values are not visible to users, but they play a critical role in how the model understands patterns, relationships, and meaning. Every time an AI system is trained, it adjusts its parameters to better predict outcomes and generate accurate responses.

Parameters exist throughout neural networks, where they act as the connections between artificial neurons. Each connection has a weight, which determines how important a particular piece of information is. During training, these weights are continuously adjusted based on data, allowing the system to improve its performance.

The Parameter Family works closely with the Data Family, which provides the information needed for learning, and the Neural Network Family, which organizes how parameters are structured and applied. It also supports the Token Family by helping the model interpret and predict sequences of language.

Understanding parameters helps students recognize that AI is not simply programmed with fixed rules. Instead, it learns by adjusting internal values that guide how it processes information. These adjustments are what allow AI systems to evolve from simple pattern recognition to complex reasoning and decision-making.

In essence, if data is what AI learns from, parameters are how AI learns.

Parameter Family Breakdown Chart

Component	Role	Description	Example
Parameters	Learning Values	Numerical values inside an AI model that are adjusted during training	Millions or billions of values inside a model
Weights	Connection Strength	Values that determine how strongly one neuron influences another	A higher weight = stronger influence
Bias	Adjustment Factor	A value added to help the model shift and fine-tune outputs	Helps improve accuracy in predictions
Training Process	Learning Mechanism	The process of adjusting parameters using data to improve performance	Model improves after analyzing many examples

Component	Role	Description	Example
Loss Function	Error Measurement	Measures how far the model's prediction is from the correct answer	Lower loss = better performance
Optimization Algorithm	Adjustment Strategy	Method used to update parameters efficiently	Gradient Descent adjusts weights step by step
Gradient	Direction of Change	Indicates how parameters should change to reduce error	Shows whether to increase or decrease a value
Backpropagation	Learning Process	System that sends error signals backward to update parameters	Adjusts earlier layers based on output error
Hyperparameters	Control Settings	External settings that guide how training happens (not learned)	Learning rate, batch size
Model Size	Capacity Indicator	The total number of parameters in a model	Larger models = more learning capacity

The Parameter Family is the hidden engine of learning in artificial intelligence. While users interact with outputs, it is the continuous adjustment of parameters behind the scenes that makes intelligent behavior possible.

Data Family Introduction

The Data Family represents the foundation of learning in artificial intelligence. Every AI system depends on data to develop knowledge, recognize patterns, and make decisions. Without data, artificial intelligence cannot learn, adapt, or function effectively.

Data can take many forms, including text, images, audio, video, and numerical information. These inputs provide the raw material that AI systems analyze during training. By studying large amounts of data, AI models begin to identify patterns, relationships, and structures that allow them to make predictions and generate responses.

The quality and diversity of data play a critical role in how well an AI system performs. Accurate, balanced, and relevant data helps models produce reliable results, while poor or biased data can lead to incorrect or unfair outcomes. This is why data collection, preparation, and evaluation are essential steps in the development of responsible AI systems.

The Data Family works closely with the Parameter Family, which adjusts internal values based on data, and the Neural Network Family, which processes and organizes data through layers of computation. It also supports the Token Family when dealing with language, as text data is converted into tokens for analysis.

Understanding the Data Family helps students recognize that AI does not "know" things on its own. Instead, it learns from the information it is given. The more meaningful and well-prepared the data, the more capable the AI system becomes.

In simple terms, data is the source of knowledge, and it is the starting point for all artificial intelligence.

Data Family Breakdown Chart

Component	Role	Description	Example
Data	Learning Material	Raw information used by AI systems to learn patterns and relationships	Text, images, audio, numbers
Dataset	Organized Collection	A structured group of data used for training and evaluation	A folder of labeled images
Training Data	Learning Input	Data used to teach the AI model during training	Thousands of sentences for a language model
Validation Data	Performance Check	Data used to tune the model during training without directly learning from it	Helps prevent overfitting
Test Data	Final Evaluation	Data used to measure how well the model performs after training	New unseen examples
Structured Data	Organized Format	Data arranged in tables with clear categories	Spreadsheets, databases
Unstructured Data	Flexible Format	Data without a predefined structure	Text documents, images, videos
Labeled Data	Guided Learning	Data tagged with correct answers to guide training	Image labeled "cat" or "dog"
Unlabeled Data	Self-Learning Input	Data without labels, used in unsupervised learning	Raw text without categories
Data Quality	Accuracy Measure	The reliability and correctness of data	Clean, complete, and consistent data
Data Bias	Imbalance Issue	When data does not represent all groups fairly	Skewed or incomplete datasets
Data Preprocessing	Preparation Step	Cleaning and organizing data before training	Removing errors, formatting text

The Data Family reminds us that artificial intelligence is only as strong as the information it learns from. High-quality data leads to meaningful insights, while poor data can limit or misguide intelligent systems.

Memory Family Introduction

The Memory Family represents how artificial intelligence systems store, retain, and retrieve information. Just as humans rely on memory to recall past experiences, learn new concepts, and make decisions, AI systems depend on memory to access previously processed information and maintain context.

In artificial intelligence, memory is not a single component but a collection of mechanisms that allow systems to handle information over time. Some forms of memory are short-term, holding information temporarily while a task is being completed. Other forms are long-term, storing learned knowledge that can be used across different tasks and interactions.

Memory plays a critical role in enabling AI systems to understand sequences, maintain context in conversations, and improve decision-making. For example, when a user interacts with a language model, the system uses memory to keep track of previous words, sentences, or prompts in order to generate coherent and relevant responses.

The Memory Family works closely with the Data Family, which provides the information to be stored, and the Parameter Family, which encodes learned knowledge within the model. It also supports the Neural Network Family by enabling systems to process sequences and retain important patterns over time.

Understanding the Memory Family helps students see that AI is not simply reacting to individual inputs in isolation. Instead, it relies on stored information and contextual awareness to produce meaningful and consistent outputs.

In simple terms, if data is what AI learns from, memory is how AI remembers and uses what it has learned.

Memory Family Breakdown Chart

Component	Role	Description	Example
Memory	Information Storage	The ability of an AI system to store and access information	Retaining previous inputs in a task
Short-Term Memory	Temporary Storage	Holds information for immediate processing	Keeping track of words in a sentence
Long-Term Memory	Persistent Storage	Stores learned knowledge for future use	Knowledge embedded in model parameters
Context Memory	Conversation Tracking	Maintains information within a specific interaction	Remembering earlier parts of a prompt
Working Memory	Active Processing	Handles information currently being used in computation	Processing a sequence step by step

Component	Role	Description	Example
External Memory	Extended Storage	Memory stored outside the model for retrieval	Databases, vector stores
Internal Memory	Embedded Knowledge	Information stored within the model itself	Learned patterns inside neural networks
Sequence Memory	Order Awareness	Tracks the order of inputs over time	Understanding sentence structure
Retrieval Mechanism	Access System	Retrieves stored information when needed	Searching relevant past data
Attention Mechanism	Focus System	Highlights important parts of memory for processing	Focusing on key words in a sentence
Memory Capacity	Storage Limit	The amount of information the system can handle at once	Context window size

The Memory Family shows that intelligence is not just about processing information in the moment, but about retaining, organizing, and using knowledge over time to create meaningful understanding.

Knowledge Family Introduction

The Knowledge Family represents how artificial intelligence organizes, connects, and uses information to create understanding. While data provides raw information and memory stores it, knowledge is what gives that information meaning and structure.

In artificial intelligence, knowledge is not simply a collection of facts. It is the result of identifying relationships, patterns, and connections within data. AI systems use knowledge to recognize concepts, make decisions, and generate meaningful responses. This allows machines to move beyond simple data processing and toward more intelligent behavior.

Knowledge can be represented in different ways within AI systems. Some forms are structured, such as databases and knowledge graphs, where relationships between concepts are clearly defined. Other forms are learned implicitly through neural networks, where patterns are encoded within parameters and used to guide predictions.

The Knowledge Family works closely with the Data Family, which provides the information needed to build knowledge, and the Memory Family, which stores and retrieves that information. It also connects with the Parameter Family, where learned knowledge is embedded within the model, and the Neural Network Family, which processes and organizes that knowledge.

Understanding the Knowledge Family helps students recognize that AI systems do not simply memorize information. Instead, they learn how pieces of information relate to one another, allowing them to interpret meaning, solve problems, and respond intelligently.

In simple terms, if data is information and memory stores it, knowledge is the understanding that makes it useful.

Knowledge Family Breakdown Chart

Component	Role	Description	Example
Knowledge	Understanding System	Organized information that allows AI to interpret meaning and make decisions	Recognizing relationships between concepts
Facts	Basic Information	Individual pieces of information stored within a system	"Water freezes at 0°C"
Relationships	Connection Mapping	Links between pieces of information that create meaning	"Teacher teaches student"
Knowledge Representation	Structure System	Methods used to organize and store knowledge in AI	Knowledge graphs, semantic networks
Knowledge Graph	Network Mapping	A structured representation of entities and their relationships	Google Knowledge Graph
Semantic Understanding	Meaning Interpretation	The ability to understand meaning behind words and data	Understanding synonyms and context
Inference	Reasoning Process	Drawing conclusions based on known information	If A = B and B = C, then A = C
Rules	Decision Framework	Logical guidelines used to make decisions	If temperature < 0°C → freeze
Ontology	Concept Framework	A structured system defining categories and relationships	Classification of animals
Explicit Knowledge	Direct Information	Clearly defined and stored knowledge	Rules, databases
Implicit Knowledge	Learned Patterns	Knowledge learned through experience and data	Patterns learned by neural networks
Knowledge Integration	System Coordination	Combining information from multiple sources	Merging text, images, and data

The Knowledge Family transforms information into understanding, allowing artificial intelligence to move beyond data and memory into meaningful reasoning and intelligent decision-making.

AI Model Family Introduction

The AI Model Family represents the core system within artificial intelligence that performs tasks such as prediction, classification, and content generation. It is the part of AI that users interact with, whether they are asking questions, generating images, or analyzing data.

An AI model is a trained system that has learned patterns from data. During training, the model processes large amounts of information and adjusts its internal parameters to improve accuracy. Once trained, the model can take new input and produce meaningful output based on what it has learned.

AI models can take many forms depending on their purpose. Some models are designed to recognize images, others to understand language, and others to make decisions or predictions. Large Language Models (LLMs), for example, are designed to process and generate human language by predicting sequences of tokens.

The AI Model Family works closely with the Data Family, which provides the learning material, and the Parameter Family, which stores learned patterns. It also relies on the Neural Network Family for structure and the Chip and Infrastructure Families for execution and deployment.

Understanding the AI Model Family helps students see that AI is not just a concept, but a working system that has been trained to perform specific tasks. It is the engine that transforms input into output.

In simple terms, if AI is the system, the model is the part that actually does the thinking and producing.

AI Model Family Breakdown Chart

Component	Role	Description	Example
AI Model	Core System	The trained system that performs tasks	Chatbot, image generator
Training	Learning Process	Process of teaching the model using data	Learning from datasets
Inference	Output Generation	Using the trained model to produce results	Answering a question
Model Architecture	Structural Design	Defines how the model is built	Neural networks, transformers
Parameters	Learned Values	Internal values adjusted during training	Billions of weights
Input Data	Entry Information	Data provided to the model for processing	User prompt
Output Data	Result	The response or prediction generated	AI-generated text
Model Types	Functional Categories	Different kinds of models for different tasks	Classification, generation
Fine-Tuning	Model Adjustment	Improving a model for specific tasks	Custom-trained AI

Component	Role	Description	Example
Pretrained Model	Base System	A model trained on large datasets before use	General-purpose AI
Evaluation	Performance Check	Measuring how well the model performs	Accuracy, loss metrics

The AI Model Family represents the working core of artificial intelligence, where learned patterns are transformed into meaningful actions, predictions, and responses.

Neural Network Family Introduction

The Neural Network Family represents the structural foundation of how artificial intelligence learns and processes information. Inspired by the human brain, neural networks are systems of interconnected units, often called neurons, that work together to analyze data, recognize patterns, and make decisions.

In artificial intelligence, a neural network is made up of layers. The input layer receives data, the hidden layers process that data, and the output layer produces a result. As information flows through these layers, the network applies mathematical transformations that allow it to detect patterns and relationships within the data.

Neural networks rely heavily on the Parameter Family, where weights and biases determine how strongly information flows between neurons. During training, these parameters are adjusted to improve the network's performance. The more the network learns, the better it becomes at recognizing patterns and making accurate predictions.

The Neural Network Family works closely with the Data Family, which provides the information needed for learning, and the Memory and Knowledge Families, which help store and organize learned information. It also supports the Token Family when processing language, allowing AI systems to interpret sequences of tokens in meaningful ways.

Understanding neural networks helps students see that AI is not simply following instructions. Instead, it is learning from examples and improving over time through layered processing and pattern recognition.

In simple terms, neural networks are the systems that allow AI to think, learn, and make sense of information.

Neural Network Family Breakdown Chart

Component	Role	Description	Example
Neural Network	Learning Structure	A system of interconnected neurons that processes data and learns patterns	Image recognition system
Neuron (Node)	Processing Unit	A basic unit that receives input, processes it, and passes output forward	A single calculation point
Input Layer	Data Entry Point	The first layer that receives raw data	Pixels of an image
Hidden Layers	Processing Layers	Intermediate layers that transform data and detect patterns	Feature detection layers
Output Layer	Result Generator	The final layer that produces predictions or decisions	"Cat" vs "Dog" classification
Weights	Signal Strength	Values that determine how strongly inputs influence outputs	Higher weight = stronger impact
Bias	Adjustment Value	Helps shift outputs to improve learning accuracy	Fine-tuning predictions
Activation Function	Decision Function	Determines whether a neuron should activate based on input	ReLU, Sigmoid
Forward Propagation	Data Flow	The process of passing input data through the network to produce output	Input → Hidden → Output
Backpropagation	Learning Process	Adjusts weights and biases based on error to improve performance	Correcting mistakes
Loss Function	Error Measurement	Calculates how far the output is from the correct answer	Prediction vs actual
Deep Neural Network	Advanced Structure	A network with many hidden layers for complex learning	Deep learning models

The Neural Network Family provides the structure that allows artificial intelligence to transform data into understanding, making it possible for machines to learn from experience and improve over time.

Dual Language System in Artificial Intelligence

Definition

The Dual Language System in Artificial Intelligence refers to the way AI communicates using two different forms of language at the same time: human language for interaction and machine language for internal processing.

Key Components

Human Language (Natural Language)
This is the language people use to communicate with AI, such as English, Spanish, Tongan, or Fijian. It includes words, sentences, and questions that users type or speak.

Machine Language (Numerical Representation)
This is how AI understands information internally. Words are converted into numbers called tokens and processed using mathematical models and algorithms.

How It Works (Step-by-Step)

A user enters a question in human language

The AI converts the words into numerical tokens

The AI processes the tokens using a neural network

The AI generates a response based on patterns and probabilities

The response is converted back into human language

Example

User Input:
"What is Artificial Intelligence?"

AI Process:

Converts words into numbers

Analyzes patterns from training data

Predicts the best possible answer

Output:
"Artificial Intelligence is the ability of machines to perform tasks that normally require human intelligence."

Why This Matters

Understanding the dual language system helps students realize that:

AI does not "think" like humans

AI translates language into math and back into language

Every interaction with AI is based on patterns, not emotions or understanding

Applications

Chatbots and virtual assistants

Language translation systems

Search engines

Voice recognition tools

Challenges

Misinterpretation of human language

Bias in language data

Lack of true understanding or reasoning

Summary

Artificial Intelligence works by translating human language into machine-readable data and then converting the results back into human language. This dual system allows AI to interact with people while operating through mathematical processes behind the scenes.

AI Translator Architecture Family Introduction

The AI Translator Architecture Family represents the system that transforms input into meaningful output in artificial intelligence. It acts as the bridge between what humans provide and what AI produces. Whether a user types a question, uploads an image, or speaks a command, this architecture is responsible for interpreting that input and generating a response.

In artificial intelligence, translation does not only mean converting one language to another. It refers to the broader process of transforming one form of information into another. For example, text can be translated into a response, speech can be converted into text, and images can be interpreted into descriptions. This transformation process is at the core of how modern AI systems function.

The AI Translator Architecture is built on multiple foundational families. It relies on the Token Family to break input into manageable units, the Neural Network Family to process patterns, and the Parameter Family to guide learning. It also uses knowledge and memory to maintain context and produce coherent results.

At the center of this architecture is the ability to encode and decode information. Encoding transforms input into a format the model can understand, while decoding converts processed information back into human-readable output. This continuous transformation allows AI systems to interact with users in a natural and meaningful way.

Understanding the AI Translator Architecture Family helps students see that AI is not simply responding randomly. Instead, it is systematically translating input into output through structured processes and learned patterns.

In simple terms, if tokens are the language pieces and neural networks are the brain, the AI Translator Architecture is the system that turns understanding into communication.

AI Translator Architecture Family Breakdown Chart

Component	Role	Description	Example
AI Translator Architecture	Transformation System	Converts input into meaningful output	Question → Answer
Input Processing	Data Intake	Receives and prepares user input for analysis	Text prompt entered by a user
Encoding	Input Conversion	Transforms input into numerical representations	Words → token IDs
Context Building	Meaning Formation	Organizes input into a structured understanding	Sentence context tracking
Sequence Processing	Order Handling	Maintains the correct order of information	Word sequence in a sentence
Attention Mechanism	Focus System	Identifies important parts of input data	Highlighting key words
Representation Layer	Feature Mapping	Converts data into internal representations for processing	Embeddings
Decoding	Output Generation	Converts processed data back into human-readable form	Token IDs → words
Output Generation	Response Creation	Produces the final answer or result	AI-generated response
Multimodal Translation	Cross-Format Processing	Translates between different types of data	Image → text description
Feedback Loop	Improvement Cycle	Uses results to refine future outputs	Model learning over time

The AI Translator Architecture Family shows that artificial intelligence is not just about understanding information, but about transforming it into meaningful communication that humans can use and interact with.

AI Architecture Family Introduction

The AI Architecture Family represents the overall design and structure of an artificial intelligence system. While individual families such as Data, Parameters, and Neural Networks explain how AI learns and processes information, the AI Architecture Family shows how all these components are organized and work together as a complete system.

In artificial intelligence, architecture refers to the blueprint that defines how data flows through a system, how components interact, and how decisions are produced. It determines how input is received, how it is processed, and how output is generated. A well-designed architecture allows AI systems to operate efficiently, scale to large problems, and produce reliable results.

AI architectures can vary depending on the type of system being built. Some are designed for image recognition, others for language processing, and others for decision-making or automation. Despite these differences, most architectures share common elements such as input layers, processing layers, memory components, and output mechanisms.

The AI Architecture Family connects all foundational families. It integrates the Data Family for input, the Token Family for language processing, the Neural Network Family for computation, the Parameter Family for learning, and the Memory and Knowledge Families for storing and organizing information. It also works closely with the AI Translator Architecture Family to transform input into meaningful output.

Understanding the AI Architecture Family helps students see the "big picture" of artificial intelligence. Instead of viewing AI as separate parts, they begin to understand it as a coordinated system where each component plays a specific role.

In simple terms, if individual families are the parts of AI, the AI Architecture Family is the blueprint that brings everything together into one working system.

AI Architecture Family Breakdown Chart

Component	Role	Description	Example
AI Architecture	System Blueprint	The overall design that defines how an AI system is structured and operates	ChatGPT system design

Component	Role	Description	Example
Input Layer	Data Entry Point	Receives raw input from users or external sources	Text prompt, image upload
Preprocessing Layer	Data Preparation	Cleans and organizes input data before processing	Tokenization, normalization
Processing Core	Computation Engine	The main system where data is analyzed and transformed	Neural networks, transformers
Parameter System	Learning Core	Stores adjustable values that guide learning and predictions	Model weights and biases
Memory System	Context Storage	Maintains and retrieves information during processing	Context window, external memory
Knowledge System	Understanding Layer	Organizes relationships and meaning within the system	Knowledge graphs, learned patterns
Decision Layer	Output Logic	Determines the final output based on processed data	Selecting the best response
Output Layer	Result Delivery	Presents results to the user in a usable format	Text, image, or audio response
Feedback Loop	Improvement System	Uses outcomes to refine future performance	Model updates and retraining
Scalability System	Expansion Capability	Allows the system to handle increasing data and complexity	Cloud-based AI systems
Integration Layer	System Connection	Connects AI with external tools and platforms	APIs, databases, applications

The AI Architecture Family reveals that artificial intelligence is not a single technology, but a coordinated system of interconnected components working together to transform data into intelligent outcomes.

Chip Family Introduction

The Chip Family represents the physical hardware that powers artificial intelligence systems. While software components such as data, algorithms, and neural networks define how AI works, none of these systems can function without the computing chips that perform the actual calculations.

Chips are specialized electronic components designed to process information at extremely high speeds. In artificial intelligence, these chips handle the complex mathematical operations required for training models and generating outputs. Every prediction, calculation, and response produced by an AI system is executed by hardware within the Chip Family.

Different types of chips are used depending on the needs of the system. Central Processing Units (CPUs) handle general-purpose tasks, while Graphics Processing Units (GPUs) are optimized for parallel processing, making them ideal for training neural networks. More specialized chips, such as Tensor Processing Units (TPUs) and Neural Processing Units (NPUs), are designed specifically for AI workloads, enabling faster and more efficient computation.

The Chip Family works closely with the Neural Network Family, which defines the structure of computation, and the Parameter Family, which stores the values being processed. It also connects to the Infrastructure Family, where large-scale systems of chips are organized in data centers and cloud environments.

Understanding the Chip Family helps students recognize that artificial intelligence is not purely abstract. It depends on physical machines that require power, resources, and engineering to operate. These chips are the engines that bring AI models to life.

In simple terms, if AI is the intelligence, chips are the machines that make that intelligence possible.

Chip Family Breakdown Chart

Component	Role	Description	Example
Chip (Processor)	Computation Engine	Electronic hardware that performs calculations required for AI	CPU, GPU
CPU (Central Processing Unit)	General Processor	Handles a wide range of computing tasks	Running basic applications
GPU (Graphics Processing Unit)	Parallel Processor	Processes many calculations at once, ideal for AI training	Training neural networks
TPU (Tensor Processing Unit)	AI Accelerator	Specialized chip designed for machine learning tasks	Google TPU
NPU (Neural Processing Unit)	Edge AI Processor	Optimized for running AI on devices like phones	Smartphone AI features
AI Accelerator	Performance Booster	Hardware designed to speed up AI computations	Dedicated AI chips
Memory (Hardware)	Data Storage	Stores data and instructions for processing	RAM, VRAM
Parallel Processing	Speed Mechanism	Ability to perform multiple calculations simultaneously	GPUs processing thousands of operations
Throughput	Processing Capacity	Amount of data processed over time	High-performance computing systems
Latency	Response Time	Time it takes to process a request	Faster chips = lower latency
Power Consumption	Energy Usage	Amount of energy required to run the chip	High-performance GPUs use more power

Component	Role	Description	Example
Edge Devices	Local Processing	Devices that run AI locally without cloud support	Phones, smart cameras

The Chip Family reminds us that behind every intelligent system is powerful hardware performing millions or billions of calculations, turning abstract models into real-world functionality.

Watts Family Introduction

The Watts Family represents the measurement of power used by artificial intelligence systems. While the Chip Family provides the hardware that performs computations, the Watts Family explains how much energy those systems consume while operating.

A watt is a unit of power that measures the rate at which energy is used. In artificial intelligence, watts help us understand how much electricity is required to run processors, train models, and generate responses. Every time an AI system processes data, performs calculations, or produces output, it consumes power.

Different AI systems require different levels of power depending on their size and complexity. Small systems, such as those running on mobile devices, use relatively low amounts of power. In contrast, large-scale AI models operating in data centers can require significant amounts of electricity, especially during training, where billions of calculations are performed continuously.

The Watts Family works closely with the Chip Family, which determines how efficiently computations are performed, and the Infrastructure Family, where large networks of machines operate together. It also connects to the overall design of AI systems, as more efficient models can reduce power consumption while maintaining performance.

Understanding the Watts Family helps students recognize that artificial intelligence is not only about intelligence and computation, but also about energy use and resource management. As AI continues to grow, managing power efficiently becomes an important part of building responsible and sustainable systems.

In simple terms, if chips perform the work, watts measure how much power it takes to do that work.

Watts Family Breakdown Chart

Component	Role	Description	Example
Watt (W)	Power Measurement	Unit that measures the rate of energy use	A device using 100 watts
Power Consumption	Energy Usage Rate	Amount of power used during operation	AI model running on GPUs
Energy Efficiency	Performance Balance	How effectively a system uses power to perform tasks	More output with less power
High-Performance Computing	Intensive Power Use	Systems that require large amounts of power for complex tasks	AI training clusters
Idle Power	Standby Usage	Power consumed when systems are not actively processing	Servers waiting for requests
Peak Power	Maximum Usage	Highest level of power used during heavy workloads	Training large AI models
Thermal Output	Heat Generation	Heat produced as a result of power usage	Cooling systems in data centers
Cooling Systems	Temperature Control	Systems used to manage heat from high power usage	Air or liquid cooling
Power Supply	Energy Source	Provides electricity to AI systems	Electrical grid, batteries
Edge Power Usage	Local Energy Use	Power consumption on smaller devices	Smartphones, IoT devices
Data Center Power	Large-Scale Usage	Power required for large AI operations	Server farms running AI models
Sustainable Energy	Efficiency Goal	Use of renewable energy to reduce environmental impact	Solar-powered data centers

The Watts Family highlights that intelligence comes with a cost—every AI system requires power, and understanding that power is essential for building efficient and sustainable technologies.

Watts Family (AI Power Usage Scale)

Definition

The **Watts Family** represents the real-time power usage of artificial intelligence systems, measured in watts (W). It shows how much power AI systems consume at a given moment as they process data and perform computations.

Watts Scale

1 Watt
10 Watts

100 Watts
500 Watts
1,000 Watts (1 Kilowatt)
10,000 Watts (10 Kilowatts)
100,000 Watts (100 Kilowatts)
1,000,000 Watts (1 Megawatt)
10,000,000 Watts (10 Megawatts)
100,000,000+ Watts (100+ Megawatts)

Key Characteristics

Measured in real-world electrical power

Higher watt values indicate higher processing demand

Used to represent the scale of AI systems from small devices to large infrastructures

Closely connected to the Chip Family, Energy Family, and Infrastructure Family

Examples

Low watts → small devices and basic AI functions

Medium watts → computers and advanced applications

High watts → data centers and large AI systems

Summary

The **Watts Family** provides a numerical representation of how much power artificial intelligence systems use in real time, helping to illustrate the scale and intensity of AI operations.

Student Insight

Watts show how much power AI is using right now.

AI Energy Family Introduction

The AI Energy Family represents the total amount of energy consumed by artificial intelligence systems over time. While the Watts Family measures the rate at which power is used at any given moment, the Energy Family focuses on the accumulated energy required to perform tasks, run systems, and sustain operations.

In artificial intelligence, energy is used every time a model is trained, a system processes data, or a response is generated. Large-scale AI systems, especially those operating in data centers, can consume significant amounts of energy due to the continuous processing of massive datasets and complex computations.

Energy is typically measured over time using units such as kilowatt-hours (kWh), which reflect how much power has been used during a specific period. For example, running a high-performance AI system for several hours or days results in a total energy cost that goes beyond the moment-to-moment power usage measured in watts.

The AI Energy Family works closely with the Watts Family, which provides the rate of power consumption, and the Chip Family, which determines how efficiently computations are performed. It also connects to the Infrastructure Family, where large-scale systems operate continuously and require long-term energy management.

Understanding the AI Energy Family helps students recognize that artificial intelligence has real-world resource implications. It is not only about performance and speed, but also about sustainability, efficiency, and responsible use of technology.

In simple terms, if watts measure how fast energy is used, energy measures how much is used over time.

AI Energy Family Breakdown Chart

Component	Role	Description	Example
Energy	Total Consumption	The total amount of power used over a period of time	Running an AI system for hours or days
Kilowatt-hour (kWh)	Energy Measurement	Unit used to measure energy consumption over time	1 kWh = using 1,000 watts for 1 hour
Energy Usage	Consumption Tracking	Total energy required for AI operations	Training a large model
Training Energy	Learning Cost	Energy used during model training processes	Weeks of GPU usage
Inference Energy	Response Cost	Energy used when AI generates outputs	Answering a user prompt
Energy Efficiency	Optimization Goal	Reducing energy use while maintaining performance	Efficient AI models
Energy Scaling	Growth Impact	Increase in energy use as systems grow larger	Bigger models = more energy
Data Center Energy	Infrastructure Demand	Total energy used by large AI facilities	Server farms operating 24/7

Component	Role	Description	Example
Cooling Energy	Temperature Control	Energy required to cool systems and prevent overheating	Air conditioning in data centers
Renewable Energy	Sustainability Source	Use of clean energy to power AI systems	Solar, wind-powered data centers
Carbon Impact	Environmental Effect	Emissions associated with energy consumption	AI training carbon footprint
Energy Management	Resource Control	Strategies to monitor and reduce energy use	Efficient scheduling, hardware optimization

The AI Energy Family reminds us that artificial intelligence operates within the physical world, where every computation consumes energy and every system has an impact on resources and sustainability.

Energy Family (Total Power Consumption of Artificial Intelligence)

Definition

The **Energy Family** represents the total amount of electricity used by artificial intelligence systems over time. It is measured in watt-hours (Wh) or kilowatt-hours (kWh) and shows how much energy AI systems consume while running.

Energy Scale

1 Watt-hour (Wh)
10 Watt-hours (Wh)
100 Watt-hours (Wh)
500 Watt-hours (Wh)
1,000 Watt-hours (1 Kilowatt-hour, kWh)
10,000 Watt-hours (10 kWh)
100,000 Watt-hours (100 kWh)
1,000,000 Watt-hours (1 Megawatt-hour, MWh)
10,000,000 Watt-hours (10 MWh)
100,000,000+ Watt-hours (100+ MWh)

Key Characteristics

Measured over time (not instant like watts)

Represents total electricity consumption

Increases the longer AI systems run

Directly related to cost and energy usage

Closely connected to Watts Family, Chip Family, and Infrastructure Family

Examples

Low energy → short AI tasks or small devices

Medium energy → daily AI usage on computers

High energy → training AI models and running data centers

Summary

The **Energy Family** provides a numerical representation of the total electricity consumed by artificial intelligence systems over time, helping to illustrate the overall cost and impact of AI operations.

Student Insight

Energy shows how much total power AI has used over time.

AI Infrastructure Family Introduction

The AI Infrastructure Family represents the large-scale systems and environments that support, power, and connect artificial intelligence technologies. While individual components such as chips, data, and models explain how AI works, infrastructure explains where and how these systems operate in the real world.

AI infrastructure includes data centers, cloud platforms, networking systems, and storage environments that allow artificial intelligence to function at scale. These systems provide the computing power, data access, and connectivity needed to train models, process information, and deliver results to users around the world.

Modern AI systems rely heavily on cloud-based infrastructure, where thousands of machines work together to handle massive workloads. This allows AI to scale efficiently, making it possible to process large datasets, run complex models, and serve millions of users simultaneously. Infrastructure also ensures reliability, security, and continuous availability of AI services.

The AI Infrastructure Family works closely with the Chip Family, which provides the hardware, the Watts and Energy Families, which manage power consumption, and the Data Family, which supplies the information being processed. It also supports the AI Architecture Family by providing the environment where all system components are deployed and connected.

Understanding the AI Infrastructure Family helps students recognize that artificial intelligence is not contained within a single device. Instead, it operates across global systems that require coordination, resources, and engineering to function effectively.

In simple terms, if AI is the system and chips are the engines, infrastructure is the environment that allows everything to run at scale.

AI Infrastructure Family Breakdown Chart

Component	Role	Description	Example
AI Infrastructure	System Environment	The physical and digital systems that support AI operations	Global AI platforms
Data Center	Processing Hub	Facilities that house servers and computing equipment	Large server farms
Cloud Computing	Scalable Platform	Remote systems that provide computing resources over the internet	AWS, Azure, Google Cloud
Servers	Compute Units	Machines that process data and run AI models	Rack-mounted servers
Networking	Connectivity System	Systems that allow communication between machines	Internet, fiber networks
Storage Systems	Data Management	Systems that store large volumes of data	Databases, cloud storage
Distributed Computing	Workload Sharing	Splitting tasks across multiple machines for efficiency	Parallel processing across clusters
Edge Infrastructure	Local Processing	Running AI closer to the user or device	Smart devices, IoT systems
Load Balancing	Traffic Control	Distributes workloads to prevent overload	Managing user requests
Security Systems	Protection Layer	Safeguards data and systems from threats	Encryption, firewalls
Redundancy	Reliability System	Backup systems to ensure continuous operation	Failover servers
Scalability	Growth Capability	Ability to expand resources as demand increases	Adding more servers dynamically

The AI Infrastructure Family reveals that artificial intelligence operates on a global scale, relying on interconnected systems that provide the power, storage, and connectivity needed to support intelligent technologies.

AI Agent Family Introduction

The AI Agent Family represents systems that can take action, make decisions, and perform tasks on behalf of users or other systems. Unlike traditional AI models that only respond to input, AI agents are designed to operate with a level of autonomy, allowing them to plan, execute, and adapt to achieve specific goals.

An AI agent receives input from its environment, processes that information, and takes actions based on predefined objectives or learned behaviors. These actions can include answering questions, automating workflows, retrieving information, or interacting with other systems. Some agents operate in simple environments, while others function in complex systems that require continuous decision-making.

AI agents often combine multiple foundational families. They use the Data Family to gather information, the Memory and Knowledge Families to retain and understand context, and the Neural Network and Parameter Families to process and learn from interactions. They also rely on the AI Architecture and Infrastructure Families to operate reliably at scale.

Modern AI agents can work independently or as part of larger systems. For example, a customer support agent can respond to inquiries automatically, while a more advanced agent can complete multi-step tasks such as scheduling, research, or system management.

Understanding the AI Agent Family helps students see that artificial intelligence is not only about generating responses, but also about taking meaningful action in real-world scenarios.

In simple terms, if AI models think and respond, AI agents act and execute.

AI Agent Family Breakdown Chart

Component	Role	Description	Example
AI Agent	Action System	An AI system that can make decisions and perform tasks	Virtual assistant
Environment	Operating Context	The space where the agent interacts and gathers information	Web, apps, databases
Input Perception	Data Intake	Receives information from the environment	User query or sensor input

Component	Role	Description	Example
Decision Engine	Action Selection	Determines what action to take based on input	Choosing a response or task
Action Execution	Task Performance	Carries out decisions in the environment	Sending an email, retrieving data
Goal System	Objective Setting	Defines what the agent is trying to achieve	Complete a task or solve a problem
Planning Module	Strategy Builder	Breaks down tasks into steps	Multi-step problem solving
Feedback Loop	Learning Cycle	Improves performance based on outcomes	Adjusting actions over time
Autonomy Level	Independence Scale	Degree of independence in decision-making	Fully automated vs assisted
Multi-Agent System	Collaboration Network	Multiple agents working together	Coordinated AI systems

Ingredient Family Introduction

The Ingredient Family represents the essential components that come together to create an artificial intelligence system. Just as a recipe requires specific ingredients to produce a final dish, AI systems rely on a combination of foundational elements working together to function effectively.

Artificial intelligence is not built from a single component. It requires data to learn from, algorithms to guide processing, models to perform tasks, hardware to execute computations, and energy to power the entire system. Each of these elements plays a critical role, and the absence of any one component would prevent the system from operating properly.

The Ingredient Family provides a simplified way for students to understand AI as a system made up of interconnected parts. Instead of viewing AI as complex or mysterious, learners can begin to see it as something constructed from identifiable and understandable components.

This family connects directly with all other foundational families, including Data, Algorithm, Neural Network, Parameter, Chip, and Energy Families. It serves as an overview that brings these elements together into a unified perspective.

Understanding the Ingredient Family helps students recognize that artificial intelligence is built, not magical. It is the result of carefully combining the right components in the right way.

In simple terms, if AI is the final product, the Ingredient Family represents everything needed to build it.

Ingredient Family Breakdown Chart

Component	Role	Description	Example
Data	Learning Input	Provides the information AI uses to learn patterns	Text, images, audio
Algorithms	Instruction Set	Guides how data is processed and decisions are made	Sorting, classification
AI Models	Execution System	Performs tasks such as prediction or generation	Language models
Neural Networks	Processing Structure	Organizes how data is processed through layers	Deep learning models
Parameters	Learning Adjustments	Fine-tune how the model learns and makes predictions	Weights and biases
Tokens	Language Units	Breaks text into smaller pieces for processing	Words or subwords
Hardware (Chips)	Computation Engine	Executes calculations required by AI systems	GPUs, CPUs
Energy	Power Source	Supplies the energy needed to run systems	Electricity usage
Infrastructure	Support System	Provides the environment for AI to operate at scale	Cloud systems, data centers
Memory	Storage System	Stores and retrieves information for processing	Context memory
Knowledge	Understanding Layer	Represents meaning and relationships within data	Knowledge graphs

The Ingredient Family shows that artificial intelligence is built from a combination of essential components, each playing a vital role in creating intelligent systems.

The Frontend and Backend Family of Artificial Intelligence

How AI Systems Connect User Interaction to Intelligent Processing

Introduction

Every artificial intelligence system operates through two essential layers: the frontend and the backend. These layers work together to transform human interaction into intelligent responses, allowing AI systems to function smoothly and effectively.

The frontend is the part of the system that users see and interact with. It includes screens, input tools, and the display of results. Whether a user is typing a question, speaking a command, or uploading an image, all interaction begins at the frontend.

Behind the scenes, the backend serves as the processing engine of the system. It receives input from the frontend, manages data, connects to the AI model, and performs the computations needed to generate a response. The backend is where the intelligence of the system operates, using models, algorithms, and infrastructure to interpret and respond to user requests.

This visual illustrates how these two layers work together as a complete system. It shows the flow of information from user input to AI processing and back to the user as a result. By understanding this structure, students can see that AI is not a single tool, but a coordinated system of layers working together.

In simple terms, the frontend is where interaction happens, and the backend is where intelligence is created.

Every AI system has two sides: what you see and what actually does the thinking.

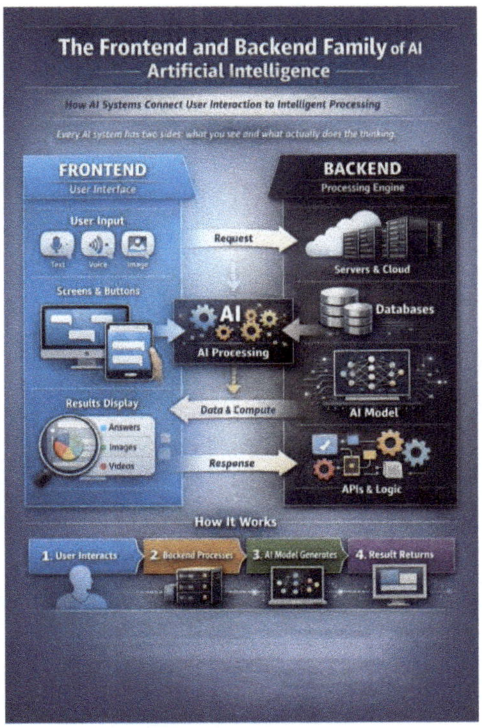

AI Qubit Family Introduction

The AI Qubit Family represents the foundation of quantum computing as it relates to artificial intelligence. While traditional computing relies on bits, which can exist as either 0 or 1, quantum computing uses qubits, which can exist in multiple states simultaneously.

A qubit is the basic unit of quantum information. Unlike classical bits, qubits can take advantage of properties such as superposition and entanglement, allowing quantum systems to process complex computations more efficiently in certain scenarios. This opens new possibilities for solving problems that are difficult or impossible for traditional computers.

In the context of artificial intelligence, qubits have the potential to accelerate learning processes, optimize large-scale systems, and improve complex simulations. Although quantum AI is still in development, it represents a future direction where computational power can expand beyond current limitations.

The AI Qubit Family connects with the Chip Family, as quantum processors are specialized hardware, and with the Algorithm Family, where new types of quantum algorithms are designed. It also relates to the Parameter and Neural Network Families, as researchers explore quantum-enhanced learning models.

Understanding the AI Qubit Family helps students see that artificial intelligence is continuously evolving. It introduces the idea that the foundations of computing themselves can change, leading to new forms of intelligence and problem-solving.

In simple terms, if bits are the foundation of today's computing, qubits represent the foundation of tomorrow's possibilities.

AI Qubit Family Breakdown Chart

Component	Role	Description	Example
Qubit	Quantum Unit	Basic unit of quantum information	Quantum bit in a quantum computer
Superposition	Multi-State Ability	Ability to exist in multiple states at once	0 and 1 simultaneously
Entanglement	Connection Property	Linking qubits so they influence each other	Paired quantum states
Quantum Gate	Operation System	Performs operations on qubits	Quantum logic gates
Quantum Circuit	Processing Structure	Sequence of quantum operations	Quantum algorithms
Quantum Processor	Hardware Engine	Specialized chip for quantum computing	Quantum computer hardware
Quantum Algorithm	Computation Method	Algorithm designed for quantum systems	Optimization problems
Quantum Speedup	Performance Advantage	Faster processing for certain problems	Complex simulations
Quantum Noise	Stability Challenge	Errors caused by environmental interference	Qubit instability
Quantum Error Correction	Stability Solution	Techniques to reduce errors in quantum systems	Fault-tolerant computing

The AI Qubit Family introduces a new frontier in computing, where the limits of traditional systems are expanded, opening the door to more powerful and advanced forms of artificial intelligence.

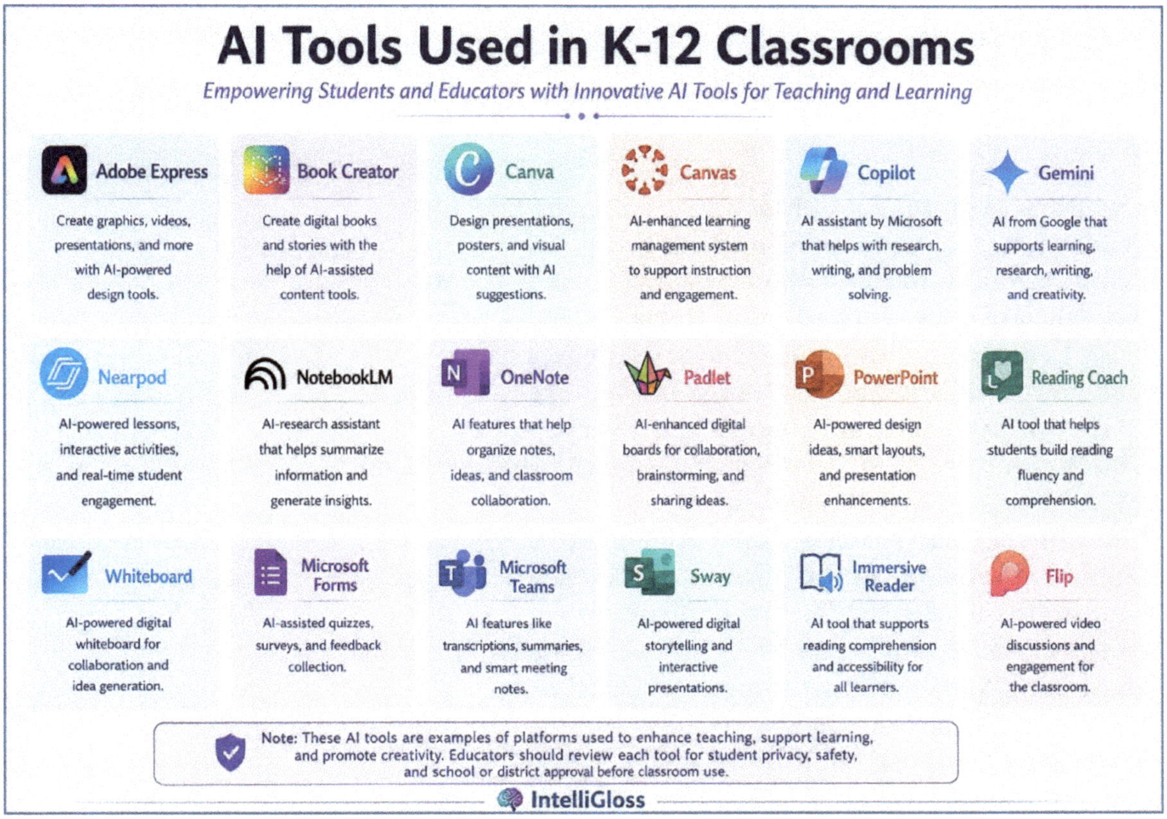

TABLE OF CONTENTS

Theme Overview

This book will show how artificial intelligence evolved from mimicking human problem-solving to *surpassing* the brightest human thinkers in logic, creativity, memory, and discovery.
It will teach students:

- How human cognition works
- How AI replicates and improves on it
- What happens when superintelligence arrives
- How humans can adapt, collaborate, and lead in a world of intelligent machines

The Future of Intelligence: Humans and Artificial Minds

Introduction — A New Chapter in Human Understanding

The twenty-first century represents one of the most transformative eras in human history. For thousands of years, intelligence was understood as a uniquely human capacity — expressed through language, creativity, scientific discovery, moral reasoning, and imagination. From ancient philosophy to modern space exploration, human intelligence shaped civilizations, cultures, and technological progress. Today, however, humanity stands at the beginning of a new chapter: the emergence of artificial minds.

Artificial Intelligence is no longer confined to science fiction or research laboratories. It assists doctors in diagnosing diseases, helps scientists analyze complex data, supports teachers in personalizing learning, guides navigation systems, translates languages in real time, and generates music, images, and written works. AI systems can process vast amounts of information at extraordinary speeds, identify patterns that humans might overlook, and perform specialized tasks with remarkable precision. These developments do not diminish human intelligence; rather, they expand the landscape in which intelligence operates.

This book explores a central question of our time: What does intelligence mean in an age where both biological and artificial systems can learn, adapt, and create? Instead of framing intelligence as a competition between humans and machines, we will examine it as a spectrum of capabilities. Human intelligence is shaped by consciousness, emotion, empathy, moral judgment, cultural understanding, and lived experience. Artificial intelligence, by contrast, is built upon data, algorithms, computational models, and engineered architectures. Each operates differently, yet both contribute meaningfully to the evolving definition of intelligence.

Throughout these pages, students and educators will explore how artificial intelligence developed from simple rule-based programs into advanced learning systems powered by neural networks and deep learning. We will examine how data fuels machine learning, how models are trained, how systems improve over time, and how computational power enables rapid innovation. At the same time, we will study the enduring qualities of human thought — creativity, ethical reasoning, adaptability, and the ability to understand context in deeply personal and cultural ways.

Understanding the future of intelligence requires balance. Artificial systems can calculate faster than humans, search enormous datasets instantly, and optimize solutions across millions of variables. Yet they do not possess lived experience, moral responsibility, or intrinsic awareness. Humans, meanwhile, may not process information at machine speed, but they bring meaning, empathy, values, and purpose to every decision. The future will not be defined by one replacing the other, but by how thoughtfully they interact.

As AI continues to influence education, healthcare, business, art, and public policy, informed understanding becomes essential. Students must learn not only how AI works, but also how to evaluate its limitations, recognize potential bias, and use it responsibly. Educators must understand how artificial systems can enhance learning without undermining critical thinking. Communities must consider how innovation can align with ethical principles and human dignity.

This book is designed to provide that foundation. It combines historical insight, scientific explanation, ethical reflection, and forward-looking analysis in language accessible to classrooms while maintaining intellectual depth. Readers will encounter key technological concepts, examine real-world applications, and engage with thoughtful questions about responsibility, creativity, and the evolving relationship between humans and machines.

The future of intelligence is not a single invention or a sudden event. It is a gradual transformation shaped by research, innovation, collaboration, and human choice. Artificial minds represent a powerful extension of human ingenuity. They reflect our curiosity, our problem-solving ability, and our desire to understand complex systems. Ultimately, the story of artificial intelligence is still a human story — one that challenges us to think carefully about what intelligence is, what it can become, and how it should be guided.

Humanity now stands at the intersection of two remarkable forms of intelligence — the biological and the artificial. Together, they will shape the next era of discovery. The question is not whether one will replace the other, but how they will coexist, collaborate, and redefine what it means to learn, create, and understand. The future of intelligence belongs to both — and its direction depends on the wisdom with which we move forward.

Chapter 1: The Birth of Human and Machine Intelligence

1.1 The Spark of Human Thought

Long before machines began to calculate or predict, the most powerful known system of intelligence was the human brain. It learned by observing the natural world, imagining possibilities, and solving the challenges of survival and cooperation. From ancient cave drawings to the construction of cities, humans demonstrated a remarkable ability to recognize patterns, communicate ideas, and make complex decisions — the essential foundations of intelligence.

What distinguishes humans is not merely strength or speed, but self-awareness — the ability to reflect, question, and think about thinking. Early philosophers such as Aristotle and Confucius explored reasoning, logic, and ethics, laying the groundwork for disciplines that would later evolve into cognitive science and artificial intelligence. Human intelligence grew through curiosity. We asked why, not just how. That desire to understand would eventually inspire the creation of machines capable of processing information in new ways.

1.2 From Human Curiosity to Mechanical Creation

As civilizations expanded, humans designed tools to extend both physical and mental capabilities. Around 500 B.C., the abacus emerged as one of the earliest devices to assist with calculation. Though simple, it allowed numbers to be processed more efficiently than mental arithmetic alone.

Centuries later, innovators such as Blaise Pascal and Gottfried Wilhelm Leibniz developed mechanical calculators capable of performing arithmetic automatically. These devices were not intelligent in themselves, but they represented a breakthrough: human reasoning could be translated into mechanical procedures.

By the nineteenth century, Charles Babbage conceptualized the Analytical Engine — a programmable machine designed to perform general calculations. Ada Lovelace, who worked alongside him, recognized something extraordinary. She proposed that such machines could manipulate symbols, not just numbers, suggesting they might one day create music or analyze patterns. Her insight foreshadowed the idea that machines could process information in ways that extend beyond simple calculation.

1.3 The Birth of Modern Computing

The twentieth century transformed these early ideas into reality. Electricity, logic circuits, and binary code gave rise to programmable computers. In the 1940s, mathematician Alan Turing posed a groundbreaking question: Can machines simulate intelligent behavior?

Turing suggested that if a machine could respond to questions in a way indistinguishable from a human, it might reasonably be considered intelligent in its behavior. His proposal, later known as the Turing Test, reshaped scientific inquiry and laid the intellectual foundation for artificial intelligence.

Early computers such as ENIAC and UNIVAC were enormous and limited in function, yet they proved something revolutionary: complex logical processes could be automated. Human reasoning could be modeled through mathematical rules and computation.

1.4 The Birth of Artificial Intelligence

In 1956, a group of scientists gathered at Dartmouth College to explore the possibility that aspects of human intelligence could be described precisely enough to be simulated by machines. This workshop marked the formal beginning of Artificial Intelligence as a scientific field.

Researchers including John McCarthy, Marvin Minsky, Allen Newell, and Herbert Simon developed early programs capable of solving logical problems and playing strategic games. These systems were limited, but they demonstrated that machines could follow structured reasoning patterns.

The vision was ambitious: to understand intelligence deeply enough to recreate certain functions through engineering.

1.5 The Human Brain as Inspiration

As researchers advanced, they turned to biology for guidance. The human brain consists of billions of neurons connected through intricate networks. Each neuron transmits signals, contributing to perception, memory, and reasoning.

Inspired by this structure, engineers developed artificial neural networks. In these systems, digital "neurons" process numerical inputs and pass results forward through layers. Through repeated exposure to data, these networks adjust internal parameters, gradually improving performance.

Unlike early rule-based systems, machine learning models can adapt based on experience. They do not possess awareness or emotion, but they can refine predictions through repetition and feedback. This shift marked a major advancement in artificial intelligence development.

1.6 Intelligence in Two Forms

By the late twentieth century, humanity had developed two distinct forms of intelligence:

• Biological intelligence — shaped by evolution, experience, emotion, culture, and consciousness.
• Artificial intelligence — engineered systems designed to process data, recognize patterns, and optimize decisions.

These forms operate differently. Human intelligence integrates emotion, moral reasoning, creativity, and lived experience. Artificial intelligence excels at rapid computation, large-scale pattern recognition, and consistent task execution.

Rather than replacing one another, these intelligences increasingly interact. Machines assist physicians in detecting diseases. Algorithms help scientists analyze climate data. AI tools support artists and educators in creative work. The relationship is evolving from simple automation to collaborative augmentation.

1.7 From Tools to Advanced Systems

As computing power increased, artificial systems began solving problems once thought to require uniquely human expertise. In 1997, IBM's Deep Blue defeated chess champion Garry

Kasparov. In 2016, AlphaGo triumphed over Go champion Lee Sedol — a milestone many experts once considered decades away.

These achievements demonstrated the power of computational strategy and machine learning. However, they also revealed an important distinction: AI systems excel within defined environments and clear rules. They do not possess understanding in the human sense. Their strength lies in specialized optimization, not broad awareness.

1.8 The Emerging Era of Coexistence

Today, artificial intelligence influences nearly every sector of society. It assists in medical research, supports climate modeling, enhances communication tools, and accelerates scientific discovery. Modern systems can generate text, analyze images, and design complex molecular structures.

Yet intelligence remains multidimensional. Machines process data with extraordinary speed and scale. Humans interpret meaning, navigate ethics, and assign purpose. The future of intelligence will not be determined by competition alone, but by how thoughtfully these capabilities are integrated.

As society enters this new era, an important question emerges:

How will humans guide the development of artificial minds to reflect shared values, responsibility, and wisdom?

The story of intelligence is no longer solely biological, nor entirely artificial. It is a shared narrative — one in which humanity continues to shape the tools it creates, and in doing so, reshapes its understanding of intelligence itself.

Chapter 1 Summary

Chapter 1 explored the historical development of intelligence in two forms: biological and artificial. It began by examining the evolution of human thought, emphasizing self-awareness, curiosity, and philosophical inquiry as foundational elements of intelligence. Human intelligence developed through observation, reasoning, and cultural transmission, forming the basis of civilization and scientific progress.

The chapter then traced the transformation of curiosity into mechanical innovation. Early calculating tools such as the abacus, followed by mechanical calculators built by Pascal and Leibniz, demonstrated that logical processes could be translated into physical mechanisms. The vision of Charles Babbage and Ada Lovelace introduced the concept of programmable machines capable of manipulating symbols, laying conceptual groundwork for computing.

The twentieth century marked a turning point with the development of electronic computers and the work of Alan Turing, who proposed that intelligent behavior could be simulated through computation. The 1956 Dartmouth Conference formally established Artificial Intelligence as a field of study, launching decades of research into machine reasoning and learning.

Inspired by the human brain, scientists developed artificial neural networks, enabling machines to learn from data rather than relying solely on fixed rules. The chapter concluded by examining the coexistence of human and artificial intelligence, highlighting both their differences and their complementary strengths.

Ultimately, intelligence today exists in two forms — biological and engineered — and the future depends on how responsibly these systems are developed and integrated.

Reflection Section

Answer the following questions in complete sentences. These prompts are designed to encourage deeper thinking and classroom discussion.

1. What characteristics make human intelligence unique compared to artificial intelligence?
2. Why was the invention of early calculating tools important for the development of AI?
3. How did Alan Turing's ideas change the scientific understanding of machine intelligence?
4. In what ways do artificial neural networks resemble the human brain? In what ways are they different?
5. Do you believe artificial intelligence should be viewed as a competitor or collaborator to human intelligence? Explain your reasoning.
6. How might society benefit from understanding both the strengths and limitations of AI?

Worksheet

Part A: Vocabulary and Concepts

Define the following terms in your own words:

1. Biological Intelligence
2. Artificial Intelligence
3. Turing Test
4. Artificial Neural Network
5. Machine Learning
6. Computational Simulation

Part B: Timeline Activity

Place the following events in chronological order and briefly describe their significance:

- The invention of the abacus
- Pascal's mechanical calculator
- Babbage's Analytical Engine

- Alan Turing's proposal of machine intelligence
- The Dartmouth Conference (1956)
- The development of artificial neural networks

Part C: Compare and Contrast

Create a short comparison between biological intelligence and artificial intelligence using at least three differences and two similarities.

Chapter 1 Quiz

Multiple Choice

1. What distinguishes human intelligence from artificial intelligence?
   A. Speed of calculation
   B. Ability to use electricity
   C. Conscious awareness and lived experience
   D. Ability to process numbers
2. The Dartmouth Conference is significant because it:
   A. Invented the first computer
   B. Formally established Artificial Intelligence as a field
   C. Created the first neural network
   D. Developed the internet
3. Alan Turing is best known for:
   A. Inventing the abacus
   B. Proposing a test to evaluate machine intelligence
   C. Building the first robot
   D. Designing microchips
4. Artificial neural networks are inspired by:
   A. Mathematical equations only
   B. The human brain's structure
   C. Mechanical gears
   D. Ancient philosophy
5. Early mechanical calculators demonstrated that:
   A. Machines could think independently
   B. Human reasoning could be translated into mechanical processes
   C. Machines had emotions
   D. Artificial intelligence had already been achieved

Short Answer

6. Explain why artificial intelligence excels at pattern recognition but lacks human awareness.
7. Describe one way AI and humans can collaborate effectively.

Answer Key

Multiple Choice Answers

1. C
2. B
3. B
4. B
5. B

Short Answer Sample Responses

6. Artificial intelligence excels at pattern recognition because it can process large amounts of data quickly and identify statistical relationships. However, it lacks awareness, consciousness, and lived experience, which are essential aspects of human understanding.
7. AI and humans can collaborate in healthcare, where AI analyzes medical images for patterns while doctors interpret results, consider patient history, and make ethical decisions.

Chapter 2: What Makes the Human Mind Unique

2.1 The Extraordinary System Inside Us

Every human being carries within them one of the most complex systems known to science — the brain.

Weighing about three pounds, it composes music, solves equations, paints masterpieces, dreams of distant futures, and invents technologies that reshape civilization. Artificial intelligence may now perform extraordinary tasks, but it was human intelligence that conceived, designed, and refined it. Understanding what makes the human mind unique is essential to understanding how artificial intelligence compares with — and differs from — its creator.

The brain contains approximately 86 billion neurons, each forming thousands of synaptic connections. These connections allow electrical and chemical signals to travel rapidly, creating dynamic networks responsible for memory, imagination, reasoning, and self-reflection — a property known as metacognition.

While AI systems can model certain reasoning processes, the human ability to reflect on one's own thoughts remains a defining characteristic. Humans can ask not only *what* they think, but *why* they think it — a layer of awareness that current artificial systems simulate through computation but do not experience as conscious reflection.

2.2 Emotion: The Engine of Meaning

Emotions are not merely feelings — they shape perception, motivation, and decision-making.

Compassion influences moral choices. Curiosity drives scientific discovery. Inspiration fuels artistic expression. Emotion and logic are deeply intertwined in human cognition, forming a system in which reasoning is rarely separate from feeling.

Artificial intelligence can analyze emotional signals through voice patterns, facial recognition, and language cues. It can generate responses that reflect empathy in structure and tone. However, these outputs arise from learned statistical associations, not lived emotional experience. AI models process representations of emotion; they do not experience sensation or internal states.

This integration of feeling and reasoning gives human intelligence a moral and relational depth that remains distinct from algorithmic processing.

2.3 Creativity: Beyond Pattern Recognition

Human creativity has shaped civilization — from early cave art to space exploration. Creativity involves connecting distant ideas, imagining possibilities that do not yet exist, and reshaping meaning through perspective.

Artificial intelligence systems can generate poems, images, music, and designs. They do so by analyzing vast datasets and recombining learned patterns in novel ways. This form of generative capability can be impressive and productive.

However, human creativity is influenced by personal memory, emotion, culture, and lived experience. While AI can produce creative outputs, the human creative process includes intentionality — the awareness of purpose and meaning behind creation.

Both human and artificial systems can generate novelty, but they do so through fundamentally different mechanisms.

2.4 Intuition: Rapid Contextual Judgment

Intuition allows humans to make rapid judgments without step-by-step analysis. A firefighter senses structural instability. A teacher perceives subtle shifts in a classroom's energy. A physician detects patterns based on years of experience.

This intuitive capacity emerges from subconscious pattern recognition shaped by lived experience. It blends memory, sensory input, and emotional awareness.

Artificial intelligence also relies on pattern recognition — often at scales far beyond human capability. However, AI requires structured training data and operates within defined parameters. Human intuition, by contrast, is deeply contextual and adaptable across varied environments.

Intuition reflects the integration of cognitive and emotional systems working together in real time.

2.5 Consciousness: The Inner Experience

One of the most profound aspects of the human mind is consciousness — the awareness of one's own existence.

Humans experience thoughts, sensations, memories, and emotions as part of an internal narrative. Neuroscientists continue to study how consciousness emerges from neural activity, and many questions remain open.

Artificial intelligence systems process inputs and generate outputs, but there is no scientific evidence that they possess subjective awareness or internal experience. Their operations are computational rather than experiential.

Distinguishing between simulation and experience remains central to understanding the difference between artificial processing and human awareness.

2.6 Morality and Ethical Reasoning

Human intelligence includes moral reasoning — the ability to evaluate right and wrong based on empathy, social values, and ethical principles.

Doctors consider patient well-being. Leaders weigh the consequences of policy decisions. Individuals reflect on fairness and justice.

AI systems can be programmed with ethical guidelines or trained on examples of acceptable behavior. However, they do not independently possess moral agency. They follow encoded objectives and learned patterns.

This distinction underscores the importance of human oversight in the development and deployment of artificial systems.

2.7 Adaptability and Learning Efficiency

Humans demonstrate remarkable adaptability. Children acquire language rapidly. Adults can learn new skills after limited exposure. The brain's plasticity allows it to reorganize and strengthen neural connections through experience.

Artificial intelligence systems also adapt through training processes. However, they often require extensive datasets and computational resources to achieve high performance.

Human learning integrates context, emotion, memory, and social interaction. AI learning relies on statistical optimization within mathematical frameworks. Both systems learn — but through distinct pathways.

2.8 Language and Meaning

Language allows humans to share ideas, preserve culture, and transmit knowledge across generations. Words carry emotional weight, historical context, and personal association.

AI language models can process enormous amounts of text and generate coherent responses. They analyze patterns in syntax and semantics to produce meaningful outputs.

Yet meaning, for humans, is connected to lived experience. When people speak, they draw upon identity, relationships, and memory. AI systems operate symbolically, mapping patterns rather than experiencing interpretation.

Language illustrates both the power and the limits of artificial systems.

2.9 Reflection and Growth

Humans possess the ability to reflect on past actions, evaluate decisions, and change behavior based on insight. Reflection transforms mistakes into learning and experience into wisdom.

Artificial intelligence systems improve performance through mathematical adjustments such as optimization algorithms. They refine outputs by minimizing error functions.

However, this process does not involve conscious self-examination or personal growth. Human reflection integrates emotional and moral dimensions, giving it transformative depth.

2.10 The Enduring Distinction

The human mind is not merely a processor of information. It is a living system shaped by emotion, culture, memory, imagination, and moral responsibility.

Artificial intelligence represents one of humanity's most sophisticated creations. It extends analytical capacity, enhances discovery, and supports innovation across disciplines.

As we consider the future of intelligence, it becomes clear that comparison alone is insufficient. Understanding requires recognizing both difference and complementarity. Human intelligence provides purpose, direction, and ethical grounding. Artificial intelligence provides scale, speed, and computational precision.

The uniqueness of the human mind is not diminished by artificial systems. Instead, it forms the foundation upon which those systems were built — and the compass that must guide their future.

Chapter 2 Summary

Chapter 2 explored the distinctive characteristics of the human mind and examined how these qualities compare with artificial intelligence systems. The chapter began by describing the extraordinary complexity of the human brain, including its billions of neurons and vast network of synaptic connections. These biological structures give rise to memory, imagination, reasoning, and metacognition — the ability to think about one's own thinking.

The chapter then examined emotion as a defining feature of human cognition. Human intelligence blends logic and feeling, shaping moral judgment, creativity, and decision-making. While AI systems can analyze emotional signals and simulate empathetic language, they do not experience internal emotional states.

Creativity was discussed as a uniquely human process shaped by lived experience, intention, and imagination. Although AI can generate creative outputs by recombining learned patterns, human creativity involves meaning, purpose, and personal perspective.

The chapter also addressed intuition, consciousness, morality, adaptability, language, and reflection. Human intelligence integrates context, ethics, and awareness in ways that artificial systems currently do not replicate. AI operates through computation and optimization, whereas human thought includes subjective experience and moral responsibility.

Ultimately, Chapter 2 emphasized that understanding the uniqueness of the human mind is essential for responsibly guiding the development of artificial intelligence. Rather than viewing AI as a replacement for humanity, the chapter presented it as a powerful tool that must be shaped by human values and oversight.

Reflection Section

Respond to the following prompts in complete sentences. Use examples where possible.

1. Why is metacognition considered a defining feature of human intelligence?
2. How does emotion influence human decision-making?
3. In what ways does AI simulate creativity, and how does this differ from human creativity?
4. Why is consciousness difficult to define scientifically?
5. What role does morality play in distinguishing human intelligence from artificial intelligence?
6. Why is human oversight necessary in the development of AI systems?
7. How does brain plasticity contribute to adaptability?
8. Do you believe artificial intelligence will ever possess consciousness? Explain your reasoning.

Worksheet

Part A: Key Concepts

Define the following terms in your own words:

1. Metacognition
2. Neural plasticity
3. Artificial Neural Network
4. Moral agency
5. Computational optimization

6. Consciousness
7. Intuition

Part B: Compare and Contrast

Create a comparison chart listing at least four differences and three similarities between human intelligence and artificial intelligence.

Human Intelligence Artificial Intelligence

Part C: Application Scenario

Read the scenario below and answer the questions:

A hospital uses AI software to analyze medical scans. The AI system identifies potential abnormalities, but the final diagnosis is made by a physician.

1. Why is the physician's role still necessary?
2. What strengths does AI contribute in this scenario?
3. What human qualities are important in communicating the diagnosis to a patient?

Chapter 2 Quiz

Multiple Choice

1. Metacognition refers to:
   A. Emotional reaction
   B. Thinking about one's own thinking
   C. Memory storage
   D. Mechanical processing
2. AI simulates emotion primarily by:
   A. Experiencing internal feelings
   B. Using biological neurons
   C. Recognizing patterns in data
   D. Developing consciousness
3. Neural plasticity refers to:
   A. Artificial programming
   B. The brain's ability to reorganize and adapt
   C. Emotional intelligence
   D. Machine retraining
4. Human morality differs from AI processing because humans:
   A. Compute faster
   B. Possess moral agency and empathy

C. Store more data
D. Use algorithms

5. Artificial intelligence primarily improves performance through:
A. Personal reflection
B. Emotional growth
C. Mathematical optimization
D. Conscious reasoning

Short Answer

6. Explain one difference between human creativity and AI-generated creativity.
7. Why is consciousness considered a major dividing line between humans and machines?
8. Describe how intuition differs from algorithmic prediction.

Answer Key

Multiple Choice Answers

1. B
2. C
3. B
4. B
5. C

Short Answer Sample Responses

6. Human creativity is shaped by lived experience, intention, and emotion, whereas AI-generated creativity recombines patterns learned from large datasets without awareness or personal meaning.
7. Consciousness is considered a dividing line because humans experience subjective awareness and internal states, while AI systems process information computationally without evidence of self-awareness.
8. Intuition arises from subconscious integration of experience, emotion, and context, while algorithmic prediction relies on statistical analysis and predefined computational processes.

Chapter 3: From Imitation to Innovation — How AI Learns

3.1 The First Stage: Imitation

Artificial Intelligence began humbly — by **imitating human behavior**.
In its earliest days, AI could not think or reason independently. It followed *rules* — strict

instructions written by humans. For example, a chess program was told what moves were legal, how to calculate points, and how to choose a move that led to victory. This stage of AI was called **rule-based intelligence**.

Rule-based systems were like students who could only recite what they memorized but never ask *why*. They were fast but not flexible. When faced with a situation outside their rules, they failed. For AI to evolve, it needed to go beyond imitation — it needed to *learn from experience.*

3.2 The Dawn of Machine Learning

The major breakthrough came with **machine learning**, a new way of programming where machines could improve their performance automatically by analyzing data.
Instead of being told what to do, the computer was given **examples** — and it learned patterns on its own.

For instance, to teach an AI to recognize cats, scientists no longer wrote "cats have fur, whiskers, and tails." Instead, they fed the computer thousands of cat images. The AI then discovered what cats looked like by itself, identifying features common among them.
This marked the beginning of AI's journey from imitation to true **innovation**.

3.3 The Cycle of Learning

Machine learning follows a cycle very similar to how humans learn:

1. **Input:** Data is collected (text, images, or numbers).
2. **Processing:** The AI model looks for patterns and relationships in that data.
3. **Prediction:** It makes a decision or prediction based on what it has seen before.
4. **Feedback:** The prediction is compared to the correct answer.
5. **Adjustment:** The AI updates itself to make fewer mistakes next time.

This cycle repeats millions of times. Each round strengthens the AI's understanding, making it more accurate. The process is called **training**, and the data used to teach it is known as the **training dataset**.

In many ways, this is how humans learn too — through trial, error, and correction.

3.4 Neural Networks: Learning Like the Brain

Inspired by the human brain, scientists built **artificial neural networks (ANNs)** — digital systems modeled after neurons. Each node in the network processes information and passes it to the next layer.
This design allows AI to identify complex patterns like faces, voices, or languages.

In a neural network:

- The **input layer** receives data (like an image or sentence).

- The **hidden layers** transform that data through millions of small calculations.
- The **output layer** delivers a decision — such as "This is a cat" or "This sentence is positive."

As the AI processes more data, it adjusts internal weights to improve accuracy — a process known as **backpropagation**. It's similar to a student correcting mistakes after checking an answer key.

3.5 From Weak AI to Learning Systems

Early AI could only perform one narrow task — such as playing chess or detecting spam emails. This is called **Narrow AI (or Weak AI)**.
However, new learning systems began to handle multiple complex tasks simultaneously, adapting to new data without reprogramming. This was the dawn of **Strong AI** concepts — systems that could generalize knowledge beyond one situation.

When DeepMind's **AlphaGo** learned strategies never seen in human play, it was not imitating anymore — it was innovating.
Its decisions surprised even the world's best players because the machine discovered creative solutions through self-learning.

Innovation had officially entered the digital brain.

3.6 The Power of Data

For machines to learn, they need **data** — and lots of it.
Data is to AI what experience is to humans. Every photo, sentence, or sound helps an algorithm understand the world better.
The more data AI processes, the more refined its predictions become. However, quality matters as much as quantity. Biased or inaccurate data can mislead AI systems, causing unfair or incorrect outcomes.

Therefore, teaching AI is like teaching a child — the lessons (data) you provide determine the kind of intelligence it grows into.

AI researchers often say, *"Garbage in, garbage out."* If poor data enters the system, poor decisions come out.

3.7 Deep Learning: The Mind That Learns Layers

Deep learning is a specialized form of machine learning that uses multiple layers of neural networks to process data. Each layer captures a deeper level of meaning:

- The first layer might detect edges in a photo.
- The next identifies shapes.
- Deeper layers recognize objects like eyes, faces, or entire people.

This layered understanding allows AI to analyze complex problems — like medical scans or languages — with human-level precision.
Deep learning doesn't just imitate human vision or hearing; it redefines what machines can *see* and *understand.*

3.8 Reinforcement Learning: Rewarding Success

Another powerful learning method is **reinforcement learning**, where AI learns by **trial and reward** — much like humans or animals.
In this method, an agent (the AI) performs an action in an environment. If the action leads to success, the AI receives a *reward*. If it fails, it is penalized.

Through thousands of repetitions, the AI discovers the best strategies for success.
This method is how robots learn to walk, self-driving cars learn to navigate traffic, and game-playing AIs learn to win through exploration and feedback.
Reinforcement learning has given AI something close to *instinct* — the ability to adjust behavior dynamically and intelligently.

3.9 Transfer Learning: Learning Faster Like Humans

Humans can apply knowledge from one experience to another.
If you learn to ride a bicycle, you can quickly learn to ride a scooter. AI now does something similar through **transfer learning**.

An AI trained to recognize cars can use that knowledge to identify trucks or motorcycles with little extra training.
This approach saves time, energy, and data — bringing AI even closer to human-style thinking.

Transfer learning shows that intelligence is not just about learning more — it's about **learning smarter**.

3.10 The Shift from Imitation to Computational Discovery

The evolution from rule-based imitation to adaptive learning represents one of the most significant technological developments of the modern era.

What began as machines executing predefined instructions has grown into systems capable of generating novel outputs based on patterns learned from data. Today's AI systems can:

- Generate written content
- Compose music
- Assist in medical research
- Model climate systems
- Suggest engineering solutions

These outputs may appear innovative because they combine learned information in ways that were not explicitly programmed line by line. However, such innovation emerges from statistical pattern recognition, optimization processes, and large-scale computation — not from independent intention or conscious creativity.

The line between imitation and generation becomes more nuanced as systems grow more complex. AI does not originate ideas from lived experience, but it can recombine knowledge at speeds and scales beyond human capability. In doing so, it becomes a powerful collaborative partner in discovery.

Artificial intelligence does not replace the human role in innovation. Rather, it expands the tools available to human thinkers. The most significant breakthroughs of the future are likely to emerge not from machines alone, but from the interaction between human insight and artificial computation.

Chapter Reflection

Artificial intelligence started as an imitation of human behavior, but through machine learning, deep learning, and reinforcement, it became an **independent innovator**.
Students reading this chapter should realize that intelligence — whether human or artificial — grows through **experience, correction, and imagination**.
The challenge of our generation is to ensure that this innovation serves humanity's values and not just its ambitions.

Chapter 3 Summary

Chapter 3 examined how artificial intelligence evolved from simple imitation to advanced learning systems capable of generating new solutions through data-driven processes.

The chapter began by explaining rule-based systems, the earliest form of AI, which followed strict human-written instructions. These systems could perform specific tasks efficiently but lacked flexibility and adaptability. They were limited to predefined rules and could not improve without manual updates.

The development of machine learning marked a turning point. Instead of programming every rule, scientists began training computers using data. By analyzing large datasets, AI systems learned patterns and improved performance over time. This learning process involves repeated cycles of input, prediction, feedback, and adjustment — similar to how humans refine skills through practice.

Artificial neural networks further advanced AI by modeling aspects of the brain's structure. Through layered processing and backpropagation, these systems adjust internal parameters to reduce error and increase accuracy. Deep learning extended this model by adding multiple layers capable of recognizing increasingly complex patterns.

Reinforcement learning introduced a system of reward and penalty, allowing AI to optimize strategies dynamically through repeated interaction with an environment. Transfer learning enabled AI systems to apply previously learned knowledge to new tasks, increasing efficiency.

The chapter concluded by emphasizing that modern AI systems generate outputs that appear innovative because they recombine learned patterns in novel ways. However, this innovation is computational and data-driven rather than conscious or intentional. AI serves as a powerful tool that enhances human capability rather than replacing human creativity or judgment.

Reflection Section

Respond in complete sentences.

1. Why were rule-based systems limited in flexibility?
2. How does machine learning differ from traditional programming?
3. In what ways does the learning cycle in AI resemble human learning?
4. Why is data quality important in training AI systems?
5. How do neural networks process information differently from rule-based systems?
6. What is the difference between reinforcement learning and supervised learning?
7. Why is transfer learning considered an efficient method?
8. Do you think AI-generated outputs should be considered creative? Explain your reasoning using evidence from the chapter.

Worksheet

Part A: Vocabulary

Define the following terms in your own words:

1. Rule-based system
2. Machine learning
3. Training dataset
4. Neural network
5. Backpropagation
6. Deep learning
7. Reinforcement learning
8. Transfer learning

Part B: Learning Cycle Diagram

Draw and label the five stages of the AI learning cycle:

1. Input
2. Processing
3. Prediction
4. Feedback
5. Adjustment

Then explain in 4–5 sentences how repetition strengthens model accuracy.

Part C: Application Scenario

A company wants to build an AI system that detects fraudulent credit card transactions.

1. What type of learning method would likely be used (supervised, reinforcement, or transfer)?
2. Why is high-quality data important in this scenario?
3. What risks might occur if the dataset is biased?

Chapter 3 Quiz

Multiple Choice

1. Rule-based AI systems primarily operate by:
   A. Learning independently from data
   B. Following predefined instructions
   C. Experiencing emotions
   D. Developing consciousness
2. Machine learning improves performance by:
   A. Rewriting its own code manually
   B. Using electrical impulses
   C. Analyzing data and adjusting parameters
   D. Random guessing
3. Backpropagation is used to:
   A. Increase emotional awareness
   B. Reduce prediction error
   C. Generate random outputs
   D. Eliminate training data
4. Reinforcement learning is based on:
   A. Memorization only
   B. Reward and penalty systems

C. Manual programming
D. Static rule sets

5. Transfer learning allows AI systems to:
   A. Start from zero each time
   B. Forget previous knowledge
   C. Apply existing knowledge to new tasks
   D. Replace human intelligence

Short Answer

6. Explain how deep learning differs from simple neural networks.
7. Why is AI innovation described as computational rather than conscious?
8. Describe one real-world example of reinforcement learning.

Answer Key

Multiple Choice Answers

1. B
2. C
3. B
4. B
5. C

Short Answer Sample Responses

6. Deep learning uses multiple layers of neural networks to identify increasingly complex patterns, whereas simple neural networks may use fewer layers and process less abstract features.
7. AI innovation is computational because it emerges from pattern recognition, optimization, and data processing rather than awareness, intention, or lived experience.
8. Reinforcement learning is used in robotics, where a robot learns to walk by receiving rewards for balanced movement and penalties for falling.

Chapter 4: Neural Networks — Modeling Intelligence in Machines

4.1 From Biological Neurons to Mathematical Models

The human brain is composed of billions of neurons that communicate through electrical and chemical signals. Each neuron receives signals through branching structures called dendrites, processes those signals in the cell body, and transmits output through an axon to other neurons. Through vast interconnected networks, these cells support perception, memory, learning, and behavior.

Scientists in the twentieth century began asking whether aspects of this biological process could be represented mathematically. Instead of attempting to recreate the brain physically, researchers developed simplified computational models inspired by neural structure. These artificial neurons accept numerical inputs, apply mathematical functions, and produce outputs.

Artificial neural networks (ANNs) are built from these simplified units. While biological neurons operate through electrochemical processes and artificial neurons operate through numerical computation, both systems rely on networks of connections that adjust over time. The goal is not to duplicate the brain, but to design systems capable of recognizing patterns and improving performance through data-driven learning.

4.2 The Structure of a Neural Network

A neural network processes information in layers:

• **Input Layer** – Receives raw data such as images, sound waves, or text.
• **Hidden Layers** – Perform mathematical transformations on the data.
• **Output Layer** – Produces the final result, such as a classification or prediction.

Each connection between artificial neurons carries a numerical value called a weight. These weights determine how strongly one signal influences another. During training, the network adjusts these weights to reduce prediction error.

As networks increase in size and complexity, they can model increasingly sophisticated patterns within data. This layered architecture enables neural networks to approximate complex relationships between inputs and outputs.

4.3 Activation Functions: Introducing Non-Linearity

Biological neurons activate when electrical signals reach a certain threshold. Artificial neurons use mathematical functions — called activation functions — to determine how input values are transformed into output values.

Common activation functions include:

• **Sigmoid** – Produces outputs between 0 and 1, often used in probability modeling.
• **ReLU (Rectified Linear Unit)** – Allows positive values to pass while setting negative values to zero, improving computational efficiency.
• **Softmax** – Converts outputs into probability distributions for classification tasks.

Activation functions introduce non-linearity into the network. Without non-linearity, neural networks would be limited to modeling simple linear relationships. With it, they can approximate more complex real-world patterns.

4.4 Learning Through Backpropagation

For a neural network to improve, it must measure its errors and adjust accordingly. This process is known as backpropagation.

The learning cycle typically follows these steps:

1. The network generates a prediction.
2. The prediction is compared to the correct answer.
3. The difference between the two is calculated as an error.
4. The error is propagated backward through the network.
5. The weights are adjusted to reduce future error.

This process is repeated across many training examples. Over time, the network refines its internal parameters to improve performance.

Backpropagation does not imply awareness or reflection. It is a mathematical optimization method that systematically reduces error through gradient-based adjustment.

4.5 Deep Neural Networks and Layered Representation

When neural networks contain many hidden layers, they are referred to as deep neural networks (DNNs). Each layer extracts progressively more abstract features from the data.

For example, in image recognition:

• Early layers may detect edges or color contrasts.
• Intermediate layers may identify shapes or textures.
• Deeper layers may recognize more complex combinations of features.

This layered processing enables deep learning systems to form hierarchical representations of patterns within data. These representations are mathematical abstractions rather than conscious interpretations.

Deep neural networks have contributed to advances in speech recognition, image classification, translation systems, and medical imaging analysis.

4.6 Specialized Architectures: CNNs and RNNs

Different tasks require different network designs.

Convolutional Neural Networks (CNNs) are optimized for image processing. They use small filters that scan across an image to detect local patterns. This approach improves efficiency and accuracy in visual recognition tasks.

Recurrent Neural Networks (RNNs) are designed for sequential data such as speech, time series, or text. They incorporate feedback loops that allow information from previous steps to influence future predictions.

These architectures do not replicate biological perception or memory. Instead, they provide mathematical tools for handling specific types of structured data.

4.7 Hardware and Parallel Computation

Training large neural networks requires substantial computational power. Modern systems rely on hardware such as Graphics Processing Units (GPUs) and Tensor Processing Units (TPUs), which can perform thousands of calculations simultaneously.

This parallel processing capability enables deep learning systems to analyze vast datasets efficiently. Advances in hardware have played a critical role in the expansion of neural network applications.

However, increased computational capacity also raises concerns about energy consumption and environmental impact, highlighting the importance of responsible technological development.

4.8 Continuous Operation and Human Oversight

Unlike biological systems, neural networks can operate continuously once deployed. They do not require rest or recovery periods. However, they also lack awareness, emotional experience, and ethical judgment.

Because neural networks optimize outputs based on training data, human oversight remains essential. Engineers and domain experts must monitor systems to ensure fairness, safety, and reliability.

Efficiency alone does not guarantee wisdom. Responsible use depends on human guidance.

4.9 Limitations of Neural Networks

Despite their capabilities, neural networks have important limitations:

- They require large amounts of labeled data for training.
- They may struggle with tasks requiring common-sense reasoning.
- They can reflect biases present in their training data.
- They often function as “black boxes,” making internal decision processes difficult to interpret.
- They demand significant computational resources.

Human intelligence, by contrast, integrates contextual understanding, moral reasoning, and adaptability from limited examples. Neural networks excel in pattern recognition but do not possess subjective awareness or independent intent.

4.9.1 The Black Box Problem

One challenge associated with many neural networks is the **black box problem**. While a neural network may produce highly accurate predictions, the internal reasoning behind those predictions is often difficult for humans to interpret. As neural networks grow deeper and more complex, understanding exactly how specific inputs lead to specific outputs becomes increasingly challenging.

For example, a neural network may correctly identify a medical image as showing signs of disease, yet the system may not clearly explain which features of the image influenced its conclusion. This lack of transparency can make it difficult for users to fully trust or verify the system's decisions.

The black box problem is especially important in high-stakes fields such as healthcare, finance, education, criminal justice, and transportation, where decisions can significantly affect people's lives. In these areas, understanding how an AI system reaches its conclusions is often just as important as the accuracy of the conclusions themselves.

To address this challenge, researchers are developing methods in **Explainable Artificial Intelligence (XAI)**. These techniques aim to make AI systems more transparent by helping users understand the factors that contribute to a model's predictions. While significant progress has been made, improving the interpretability of neural networks remains an active area of research.

Understanding the black box problem helps students recognize that powerful AI systems are not always fully understandable. Responsible AI development requires both strong performance and a commitment to transparency, accountability, and human oversight.

4.10 Emerging Directions in Artificial Neural Systems

Researchers continue exploring new approaches inspired by neuroscience, including spiking neural networks and neuromorphic hardware, which aim to improve energy efficiency and computational design.

Other areas of research examine how advanced computing methods may enhance model complexity and performance. These developments remain experimental and focus on improving efficiency and scalability rather than replicating biological consciousness.

As artificial systems grow more sophisticated, ongoing research seeks to better understand both their capabilities and their boundaries. The study of neural networks is not an attempt to recreate the human brain in full, but to develop powerful computational tools that support human inquiry and innovation.

Chapter Reflection

Neural networks represent one of the most significant advances in modern computing. Inspired by biological structure but grounded in mathematics, they enable machines to detect patterns, classify information, and make predictions at scale.

Yet neural networks remain computational systems. They optimize numerical relationships within data but do not experience awareness, emotion, or moral judgment.

Understanding both the power and the limitations of these systems is essential. As artificial networks become more integrated into daily life, human responsibility remains central. The future of intelligence depends not only on technological progress, but on thoughtful guidance and ethical application.

Chapter 4: Neural Networks — Modeling Intelligence in Machines

This is structured for district use, academically grounded, and aligned with high school AI literacy standards.

Chapter 4 Summary

Chapter 4 explored how neural networks model aspects of learning in machines. Inspired by biological neurons, artificial neural networks (ANNs) are mathematical systems designed to recognize patterns and improve performance through data-driven training.

The chapter began by explaining the structure of biological neurons and how scientists developed simplified computational models to simulate signal processing. Artificial neurons receive numerical inputs, apply mathematical transformations, and produce outputs. When connected in layers, these units form neural networks capable of modeling complex relationships in data.

Neural networks consist of input, hidden, and output layers. Each connection carries a weight that adjusts during training to reduce error. Activation functions introduce non-linearity, enabling networks to handle complex patterns rather than simple linear relationships.

Backpropagation was introduced as the core learning mechanism, allowing networks to measure prediction errors and adjust weights accordingly. Deep neural networks expand this structure by using multiple hidden layers to form hierarchical representations of data.

Specialized architectures such as Convolutional Neural Networks (CNNs) and Recurrent Neural Networks (RNNs) were discussed as tools for handling image and sequential data. The chapter also examined the importance of parallel computation using GPUs and TPUs.

Finally, the chapter emphasized the limitations of neural networks, including data dependence, interpretability challenges, bias risks, and lack of consciousness or ethical reasoning. Neural networks are powerful computational tools, but they require human oversight and responsible implementation.

Reflection Section

Respond in complete sentences.

1. Why are artificial neural networks considered inspired by biology but not equivalent to the human brain?
2. How do activation functions improve the capability of neural networks?
3. Why is backpropagation essential for machine learning?
4. What is meant by "hierarchical representation" in deep learning?
5. Why are CNNs well-suited for image recognition?
6. What ethical responsibilities arise when deploying neural networks?
7. How does computational power influence the growth of AI capabilities?
8. Why is it important to understand the limitations of neural networks?

Worksheet

Part A: Vocabulary Definitions

Define the following terms in your own words:

1. Artificial neuron
2. Neural network
3. Weight (in machine learning)
4. Activation function
5. Backpropagation
6. Deep neural network
7. Convolutional Neural Network (CNN)
8. Recurrent Neural Network (RNN)
9. Parallel processing
10. Black box problem

Part B: Diagram Activity

Draw a simple neural network with:

- 1 input layer
- 1 hidden layer
- 1 output layer

Label:

- Inputs
- Weights

- Activation functions
- Outputs

Write 3–4 sentences explaining how data flows through the network.

Part C: Application Scenario

A hospital uses a deep neural network to analyze X-ray images.

1. Why might a CNN be appropriate for this task?
2. What role does backpropagation play during training?
3. Why should human doctors still review the AI's output?
4. What risks exist if the training data lacks diversity?

Chapter 4 Quiz

Multiple Choice

1. Artificial neural networks are designed to:
   A. Replicate human emotions
   B. Simulate aspects of pattern recognition
   C. Develop consciousness
   D. Replace biological intelligence
2. The purpose of activation functions is to:
   A. Store memory permanently
   B. Introduce non-linearity into computations
   C. Reduce computational power
   D. Eliminate training data
3. Backpropagation is used to:
   A. Randomize outputs
   B. Measure and reduce prediction error
   C. Increase emotional intelligence
   D. Create new neurons
4. Deep neural networks differ from simple neural networks because they:
   A. Use fewer layers
   B. Eliminate weights
   C. Contain multiple hidden layers
   D. Do not require training
5. Convolutional Neural Networks are especially useful for:
   A. Ethical reasoning
   B. Image and visual processing
   C. Emotional modeling
   D. Conscious awareness
6. One major limitation of neural networks is:
   A. They require no data
   B. They always explain their reasoning clearly

C. They may reflect biases in training data
D. They cannot process numbers

Short Answer

7. Explain why neural networks are described as computational rather than conscious systems.
8. Describe how GPUs contribute to modern AI development.
9. Why is human oversight essential when deploying neural network systems?

Answer Key

Multiple Choice Answers

1. B
2. B
3. B
4. C
5. B
6. C

Short Answer Sample Responses

7. Neural networks are computational systems because they process numerical data using mathematical optimization techniques. They do not possess awareness, intention, or subjective experience.
8. GPUs enable parallel processing, allowing thousands of calculations to occur simultaneously. This significantly accelerates the training of large neural networks.
9. Human oversight is essential because neural networks may produce biased or incorrect outputs. Humans are needed to interpret results, ensure ethical use, and make final decisions.

Chapter 5: Deep Learning — Expanding the Capabilities of Artificial Intelligence

5.1 From Shallow Models to Deeper Representations

Early machine-learning systems were often described as *shallow* because they used only one or two hidden layers to process information. These models could identify simple relationships in data — such as basic shapes in images or straightforward patterns in numbers — but struggled with more complex tasks involving language, facial recognition, or natural sound.

The development of **deep learning** marked a major advancement. Deep learning systems use multiple layers of artificial neurons, allowing them to build increasingly abstract representations of data. Rather than relying on manually crafted rules, these systems learn hierarchical patterns directly from large datasets.

The term "deep" refers not to intelligence or awareness, but to the number of computational layers involved. Each additional layer enables the model to transform data in more sophisticated ways, improving its ability to recognize complex structures within information.

5.2 How Deep Learning Processes Information

Deep learning systems rely on deep neural networks (DNNs) trained on extensive datasets. During training, the system adjusts millions — and sometimes billions — of parameters to reduce prediction errors.

For example, in handwriting recognition:

1. Early layers detect simple visual features such as lines or curves.
2. Intermediate layers combine these features into shapes resembling letters.
3. Later layers classify full characters or words based on learned patterns.

Importantly, the system does not "see" or "interpret" meaning in a human sense. Instead, it identifies statistical relationships within data. Through repeated training cycles, these statistical relationships form structured representations that allow the model to perform classification, prediction, or generation tasks with high accuracy.

5.3 The Role of Computational Power

The growth of deep learning has been closely linked to advances in computing hardware. Graphics Processing Units (GPUs) and Tensor Processing Units (TPUs) can perform thousands of mathematical operations simultaneously. This parallel processing capability significantly reduces training time for large models.

Cloud computing infrastructure further accelerated development by providing scalable storage and processing resources. As a result, deep learning systems became capable of analyzing enormous datasets — from astronomical images to medical records — with increasing efficiency.

However, the expansion of computational capacity also raises important considerations about energy consumption and environmental sustainability. Responsible AI development requires balancing innovation with resource awareness.

5.4 Training a Deep Learning Model

Training a deep learning system involves several structured steps:

1. **Data Preparation** – Collecting, cleaning, labeling, and organizing relevant data.
2. **Model Design** – Selecting the architecture and number of layers.
3. **Optimization** – Using algorithms such as Stochastic Gradient Descent to minimize error.
4. **Evaluation** – Testing performance on new data and adjusting to prevent overfitting.

Overfitting occurs when a model memorizes training examples rather than learning general patterns. Preventing overfitting ensures that the system performs well in real-world scenarios.

The training process is mathematical and iterative. It improves performance through systematic adjustment rather than conscious reflection.

5.5 Specialized Architectures in Deep Learning

As deep learning evolved, researchers developed specialized architectures tailored to different types of data.

Convolutional Neural Networks (CNNs) are optimized for image processing. By applying filters across visual data, they detect spatial patterns such as edges, textures, and shapes.

Recurrent Neural Networks (RNNs) are designed for sequential data, including speech, music, and time series. They incorporate feedback loops that allow earlier inputs to influence later predictions.

Transformers, a more recent architecture, use attention mechanisms to process large amounts of data in parallel. Attention mechanisms allow the system to weigh the relevance of different parts of input data when generating an output. Transformers form the foundation of many modern large language models.

These architectures improve computational efficiency and performance, but they remain mathematical systems that model patterns rather than experience understanding.

5.6 The Role of Data in Deep Learning

Deep learning depends heavily on data. Large datasets provide the examples necessary for identifying patterns and refining predictions.

Data sources today include images, audio recordings, text, sensor readings, and scientific measurements. The scale of available digital information has significantly contributed to the growth of AI capabilities.

However, data quality is as important as data quantity. Biased, incomplete, or inaccurate data can lead to unfair or unreliable outputs. Ethical AI development requires careful dataset selection, validation, and ongoing review.

The effectiveness of deep learning reflects the information it receives.

5.7 Transfer Learning and Efficiency

Transfer learning improves efficiency by allowing models trained on one task to adapt to another related task with minimal additional training.

For example, a model trained on general image recognition can be fine-tuned to detect medical anomalies or specific object categories. This reduces the need for building entirely new systems from the beginning.

Transfer learning reflects a broader principle in AI development: leveraging prior knowledge to accelerate progress. It reduces computational costs and makes advanced systems more accessible to researchers and educators.

5.8 High Performance in Specialized Domains

Deep learning systems have achieved remarkable results in specialized tasks.

In 2016, AlphaGo defeated world champion Lee Sedol in the complex board game Go. In 2020, AlphaFold accurately predicted protein structures that had challenged researchers for decades. AI systems are now used to assist in medical diagnostics, climate modeling, logistics optimization, and language translation.

These achievements demonstrate that deep learning models can achieve exceptional performance in well-defined domains. However, such systems operate within the boundaries of their training data and task objectives. Their capabilities are domain-specific rather than general.

Performance in specialized tasks does not imply broad human-like reasoning across all areas of cognition.

5.9 Ethical and Technical Challenges

As deep learning systems become more integrated into society, challenges emerge.

- **Bias** – Models may reflect inequalities present in training data.
- **Explainability** – Complex networks can be difficult to interpret.
- **Energy Consumption** – Large-scale training requires substantial computational resources.
- **Security Risks** – Systems may be vulnerable to adversarial attacks.

Addressing these challenges requires interdisciplinary collaboration between engineers, policymakers, educators, and ethicists.

Technological advancement must be guided by responsibility.

5.10 The Evolving Landscape of Artificial Intelligence

Deep learning remains an active area of research. New approaches include self-supervised learning, which reduces reliance on labeled data, and multimodal systems that integrate text, image, and audio inputs.

Some researchers explore the concept of Artificial General Intelligence (AGI) — systems capable of transferring knowledge across diverse domains. At present, AGI remains theoretical and subject to ongoing debate within the scientific community.

Deep learning has significantly expanded the practical capabilities of artificial systems, but questions about general reasoning, common sense, and consciousness remain open.

The future of AI will depend not only on technical breakthroughs but also on thoughtful governance, ethical standards, and human collaboration.

Chapter Reflection

Deep learning represents a major expansion in artificial intelligence capability. By combining layered neural architectures, large datasets, and advanced computational power, it enables systems to perform complex tasks with high accuracy.

Yet deep learning remains grounded in mathematical optimization and statistical modeling. It does not possess awareness, intention, or independent moral judgment.

Understanding both its strengths and its boundaries prepares students to engage responsibly with AI technologies. The future of intelligence will be shaped not by machines alone, but by the partnership between human insight and artificial computation.

Chapter 5: Deep Learning — Expanding the Capabilities of Artificial Intelligence

Structured for high school AI literacy, district-ready, academically grounded, and aligned with your human-centered framework.

Chapter 5 Summary

Chapter 5 examined the development and impact of deep learning as a major advancement in artificial intelligence. Early machine-learning systems, often described as shallow models, could identify basic patterns but struggled with complex tasks. Deep learning introduced multi-layered neural networks capable of forming hierarchical representations of data.

These deep neural networks (DNNs) process information through multiple hidden layers, allowing systems to model increasingly abstract relationships. Activation functions and backpropagation enable networks to reduce error and improve accuracy through repeated training cycles.

The rise of powerful hardware such as GPUs and TPUs accelerated deep learning by enabling parallel computation at large scale. This computational growth allowed models to analyze massive datasets efficiently.

Specialized architectures such as Convolutional Neural Networks (CNNs), Recurrent Neural Networks (RNNs), and Transformers expanded AI capabilities across images, sequential data, and language processing.

The chapter also addressed the importance of data quality, transfer learning for efficiency, and the ethical challenges associated with bias, interpretability, and sustainability.

While deep learning systems have achieved high performance in specialized domains, they remain computational tools that rely on statistical modeling rather than conscious reasoning. Their future development depends on responsible human oversight and ethical application.

Reflection Section

Respond in complete sentences.

1. Why were shallow machine-learning models limited in handling complex tasks?
2. What does "depth" refer to in deep learning?
3. How do hierarchical representations improve performance in deep neural networks?
4. Why is computational power essential for deep learning?
5. What role does backpropagation play in improving model accuracy?
6. How does transfer learning increase efficiency?
7. Why is data quality as important as data quantity?
8. What ethical responsibilities arise when deploying deep learning systems?
9. Why is high performance in a narrow domain different from general intelligence?
10. How can educators help students think critically about deep learning technologies?

Worksheet

Part A: Vocabulary

Define the following terms in your own words:

1. Deep learning
2. Shallow model
3. Hierarchical representation
4. GPU
5. TPU

6. Overfitting
7. Transfer learning
8. Attention mechanism
9. Multimodal learning
10. Artificial General Intelligence (AGI)

Part B: Concept Mapping

Create a concept map showing how the following elements connect:

- Data
- Neural networks
- Backpropagation
- Computational power
- Specialized architectures
- Ethical oversight

Write a short paragraph explaining how these components interact.

Part C: Application Scenario

A research lab trains a deep-learning model to detect early signs of disease in medical images.

1. Why would a CNN likely be used?
2. Why is diverse training data important?
3. What risks exist if the model overfits?
4. Why must doctors remain involved in final decisions?

Chapter 5 Quiz

Multiple Choice

1. The term “deep” in deep learning refers to:
   A. Human-level reasoning
   B. The number of computational layers in a network
   C. Emotional intelligence
   D. Speed of processing
2. Backpropagation is primarily used to:
   A. Increase energy efficiency
   B. Reduce prediction error
   C. Create consciousness
   D. Eliminate datasets
3. CNNs are particularly effective for:
   A. Ethical decision-making

B. Image processing tasks
C. Hardware manufacturing
D. Emotional modeling
4. Transfer learning allows a model to:
A. Forget previous knowledge
B. Start training from scratch
C. Apply learned features to new tasks
D. Replace human teachers
5. A major ethical concern in deep learning is:
A. Lack of internet access
B. Overly simple algorithms
C. Bias in training data
D. Too little computational power
6. High performance in a narrow domain does NOT necessarily imply:
A. Specialized capability
B. General human-like reasoning
C. Effective pattern recognition
D. Statistical optimization

Short Answer

7. Explain why deep learning systems are described as computational rather than conscious.
8. Describe one advantage and one limitation of large-scale computational power in AI.
9. Why is human oversight necessary even when deep learning systems achieve high accuracy?

Answer Key

Multiple Choice Answers

1. B
2. B
3. B
4. C
5. C
6. B

Short Answer Sample Responses

7. Deep learning systems operate by adjusting numerical weights through mathematical optimization. They do not possess awareness, intention, or subjective experience, making them computational rather than conscious systems.
8. Advantage: Large-scale computational power allows faster training and more accurate modeling of complex data.
Limitation: It requires significant energy resources and can increase environmental impact.

9. Human oversight is necessary because models can produce biased, incomplete, or incorrect outputs. Humans ensure ethical use, interpret results responsibly, and provide contextual judgment.

Chapter 6: The Power of Data and the Limits of Memory

6.1 Data — The Foundation of Artificial Intelligence

Every system that learns depends on information. For human beings, information arrives through lived experience — through observation, conversation, study, and reflection. For machines, information arrives as data: structured or unstructured collections of numbers, images, symbols, measurements, and signals.

In artificial intelligence, data is not optional. It is foundational. Algorithms do not invent knowledge from nothing; they analyze examples. Without exposure to data, an AI model cannot detect patterns, identify relationships, or improve performance.

Just as a human child develops language by hearing words repeatedly, a machine-learning model develops predictive ability by processing thousands or millions of examples. Data supplies the raw material from which artificial systems construct their internal representations of the world.

However, data alone does not create intelligence. It must be organized, interpreted, and refined. The relationship between data and intelligence is not automatic — it is structured through design.

6.2 Experience and Memory in the Human Brain

Human beings also rely on accumulated information. Every lesson learned, mistake made, or conversation held contributes to memory formation. The brain stores experience through networks of neurons that strengthen connections when repeated.

Unlike computers, human memory is dynamic. It changes over time. Memories can fade, merge, or shift in emotional intensity. While this imperfection may seem like a weakness, it serves an important purpose: it allows prioritization.

Humans do not store every detail of every moment. Instead, the brain emphasizes experiences that carry meaning — events connected to emotion, importance, or survival. Forgetting is part of cognitive efficiency.

This selective memory allows humans to transform raw experience into judgment, empathy, and wisdom. Machines, in contrast, do not prioritize based on meaning. They store according to structure.

6.3 How Artificial Systems Perceive Information

Artificial intelligence does not "see" or "hear" in the human sense. It converts sensory inputs into numerical representations.

- An image becomes a matrix of pixel values.
- A voice recording becomes a waveform represented by numerical frequencies.
- A sentence becomes tokens mapped to vectors in high-dimensional space.

Once information is digitized, the system searches for statistical patterns. For example, when recognizing a handwritten letter, the model does not understand language as humans do. It analyzes shapes, contrasts, and structural similarities.

This mathematical perception allows AI to process information at extraordinary scale. Systems can analyze entire libraries of text or millions of medical scans in a fraction of the time required by human experts.

Yet speed does not equal understanding. Artificial perception identifies correlations, not lived meaning.

6.4 The Era of Big Data

Modern society generates unprecedented volumes of digital information. Social media interactions, financial transactions, sensor networks, satellite imagery, and online communication continuously expand global data reserves.

Big Data is often described through three characteristics:

- **Volume** — Massive quantities of information
- **Velocity** — Rapid, real-time generation
- **Variety** — Multiple formats (text, video, audio, structured databases)

This abundance fuels deep learning systems. Large-scale datasets allow models to detect subtle patterns and improve accuracy.

However, dependency on large datasets introduces vulnerability. When data flows slow or quality declines, model performance suffers. Intelligence built on data requires ongoing maintenance.

6.5 Machine Memory Systems

Machines store information through digital memory architectures — including RAM, solid-state drives, cloud storage, and distributed databases. These systems allow rapid retrieval of stored values.

In addition to external storage, trained models contain internal representations: compressed mathematical encodings of learned patterns. These encodings function as a form of artificial memory.

However, this memory differs fundamentally from human recall. Artificial memory retrieves data without emotion, narrative, or context. It does not attach personal significance to events.

A machine may recall millions of patterns, but it does not remember a single experience.

Human memory is limited but meaningful. Machine memory is expansive but mechanical.

6.6 Biological Limits of Human Memory

The human brain has capacity limits. Cognitive resources such as attention and working memory restrict how much information can be processed at once.

Over time, unused information fades. Yet this fading serves adaptation. By filtering details, the brain preserves flexibility and learning efficiency.

Emotion also shapes retention. Events tied to joy, fear, or personal significance become more deeply encoded. This emotional dimension guides moral development and identity formation.

Artificial systems lack this emotional filtering. All data is weighted mathematically rather than morally.

Where humans remember what matters, machines retain what is programmed.

6.7 Technical Limits of Machine Memory

Although machines can store enormous quantities of data, they face constraints:

- Storage infrastructure requires physical resources.
- Large-scale data centers consume significant energy.
- Excessive memorization may reduce generalization ability.

In machine learning, overfitting occurs when a model memorizes training data too precisely, preventing flexible application to new situations. Effective intelligence requires abstraction, not repetition.

Engineers design systems to extract patterns rather than preserve raw detail. In this way, artificial systems attempt to mimic one strength of the human brain: storing principles instead of isolated facts.

6.8 Data Quality and Ethical Implications

Intelligence reflects its training environment. If data contains inaccuracies, bias, or imbalance, those distortions influence model outputs.

Historical examples have shown that insufficient diversity in datasets can produce uneven performance across demographic groups. These outcomes highlight the responsibility embedded in data collection and curation.

Artificial intelligence does not independently evaluate fairness. It reflects patterns within its input data.

Therefore, ethical AI development begins with ethical data practices — including transparency, representation, and careful evaluation.

Students studying AI must understand that datasets are shaped by human choices. Technology inherits those decisions.

6.9 Privacy, Ownership, and Digital Footprints

Modern individuals generate digital traces through online activity, purchases, communication, and location tracking. These traces contribute to the data ecosystems that train AI systems.

Privacy regulations such as the **General Data Protection Regulation (GDPR)** and the **California Consumer Privacy Act (CCPA)** establish rights regarding data use and access. These frameworks aim to balance innovation with personal autonomy.

As artificial intelligence expands, societies must continue addressing questions of consent, transparency, and accountability.

Data fuels intelligence — but ownership defines justice.

6.10 The Future of Data and Adaptive Memory

Emerging research explores synthetic data generation, in which AI systems create realistic but artificial datasets. This approach may reduce reliance on sensitive personal information.

Other innovations include neuromorphic hardware designed to mimic biological memory structures, emphasizing energy efficiency and adaptability.

Future systems may improve their capacity to store and retrieve information dynamically. Yet even as artificial memory evolves, it remains fundamentally computational rather than conscious.

The distinction between storage and understanding persists.

Chapter Reflection

Data empowers artificial intelligence, while memory shapes human understanding. Humans transform experience into meaning through emotion, reflection, and selective retention. Machines transform data into predictions through computation and pattern recognition.

Both systems face limits. Humans forget; machines lack interpretation. The partnership between biological insight and artificial precision defines the next stage of intelligence.

Students must recognize that information becomes knowledge only through responsible use. The future of intelligence depends not solely on data volume, but on human wisdom guiding its application.

Chapter 6: The Power of Data and the Limits of Memory

Structured for high school AI literacy, classroom-ready, academically rigorous, and aligned with your human-centered approach.

Chapter 6 Summary

Chapter 6 explored the central role of data in artificial intelligence and compared machine memory with human memory. Artificial intelligence depends entirely on data to learn patterns, make predictions, and improve performance. Without sufficient and well-structured data, AI systems cannot function effectively.

Humans also rely on information, but their memory operates differently. Human memory is selective, emotional, and adaptive. It prioritizes meaning over precision and allows individuals to grow through reflection and experience. Machines, by contrast, store and retrieve data without emotion or personal context.

The chapter examined how AI systems digitize sensory inputs into numerical form and analyze statistical relationships. It introduced the concept of Big Data, emphasizing volume, velocity, and variety as defining features of the modern information era.

Students also learned that machine memory, while vast, faces technical constraints such as energy consumption and overfitting. Meanwhile, human memory, though limited, provides depth and moral context.

Finally, the chapter addressed ethical responsibilities related to data quality, bias, privacy, and ownership. Intelligence reflects the data used to train it, making human oversight essential.

Reflection Section

Answer in complete sentences.

1. Why is data essential for artificial intelligence systems?
2. How does human memory differ from machine memory?
3. Why is forgetting considered beneficial for human learning?
4. What are the three characteristics of Big Data?
5. Why does overfitting limit a machine-learning model's effectiveness?
6. How can biased data affect AI outcomes?
7. Why is privacy an important issue in the age of AI?

8. How does emotional memory influence human decision-making?
9. Why does storing more data not automatically lead to better intelligence?
10. What responsibilities do developers have when collecting training data?

Worksheet

Part A: Vocabulary Practice

Define each term in your own words:

1. Data
2. Internal representation
3. Big Data
4. Overfitting
5. Synthetic data
6. Neuromorphic hardware
7. Data privacy
8. Digital footprint
9. Generalization
10. Ethical AI

Part B: Compare and Contrast

Create a two-column chart comparing **Human Memory** and **Machine Memory**. Include at least five differences in structure, function, and limitation.

After completing the chart, write one paragraph explaining which system you believe is more adaptable and why.

Part C: Scenario Analysis

A company develops an AI hiring system trained primarily on resumes from one geographic region.

1. What potential risks exist in this dataset?
2. How might bias appear in hiring decisions?
3. What steps could developers take to improve fairness?
4. Why is transparency important in this scenario?

Chapter 6 Quiz

Multiple Choice

1. Artificial intelligence learns primarily through:
   A. Emotional experience

B. Instinct
C. Data analysis
D. Conscious reflection

2. Human memory differs from machine memory because it:
   A. Stores unlimited information
   B. Is shaped by emotion and meaning
   C. Never forgets
   D. Operates only numerically
3. The three characteristics of Big Data include:
   A. Speed, memory, and logic
   B. Volume, velocity, and variety
   C. Depth, width, and height
   D. Storage, ethics, and bias
4. Overfitting occurs when a model:
   A. Learns patterns too broadly
   B. Fails to train
   C. Memorizes training data too precisely
   D. Uses insufficient memory
5. Biased datasets can lead to:
   A. Improved fairness
   B. Neutral predictions
   C. Unbalanced or inaccurate outcomes
   D. Reduced storage requirements
6. Digital footprints refer to:
   A. Physical storage devices
   B. Personal data trails created online
   C. Printed documents
   D. Computer hardware components

Short Answer

7. Explain why more data does not automatically guarantee better intelligence.
8. Describe one advantage and one limitation of machine memory.
9. Why must ethical considerations guide data collection in AI systems?

Answer Key

Multiple Choice

1. C
2. B
3. B
4. C
5. C
6. B

Short Answer Sample Responses

7. More data improves learning only if the data is accurate, diverse, and relevant. Poor-quality or biased data can mislead models regardless of quantity.
8. Advantage: Machine memory can store and retrieve enormous amounts of information quickly.
   Limitation: It lacks emotional meaning and may overfit or require significant energy resources.
9. Data shapes AI decisions. If collected irresponsibly or without fairness, AI systems may produce harmful or discriminatory outcomes. Ethical oversight ensures responsible innovation.

Chapter 7: Artificial General Intelligence — Expanding the Boundaries of Machine Reasoning

7.1 Beyond Specialized Intelligence

Artificial intelligence today performs remarkable tasks. It translates languages, detects diseases in medical images, recommends music, and answers complex questions. Yet each of these abilities exists within a boundary. Most current systems are designed to perform one category of task extremely well. This is known as Artificial Narrow Intelligence (ANI).

Artificial General Intelligence (AGI), however, represents a different concept. AGI refers to a theoretical form of AI capable of learning, reasoning, and adapting across a wide range of domains — not just one. Rather than being programmed or trained separately for each task, an AGI system would apply knowledge flexibly, transferring learning from one situation to another.

AGI is not yet achieved. It remains a research goal. But it represents an important question in science: Can machines one day demonstrate broad, adaptable reasoning similar to human intelligence?

7.2 Understanding Generalization

The difference between narrow and general intelligence lies in generalization.

A narrow AI trained to recognize animals in images cannot automatically write an essay about wildlife conservation. A chess-playing program cannot suddenly solve a physics equation without new training.

Humans, by contrast, apply knowledge across fields. A student who learns logical reasoning in mathematics may apply that reasoning in science, debate, or programming. This flexibility defines general intelligence.

AGI research seeks to create systems capable of similar cross-domain transfer. Instead of isolated skills, such a system would integrate knowledge into a unified model of understanding.

7.3 Components of Broad Intelligence

To approach general intelligence, a machine would need to integrate multiple capabilities, including:

- **Reasoning** — solving unfamiliar problems logically
- **Learning efficiency** — acquiring new skills with limited data
- **Context awareness** — understanding how meaning shifts across situations
- **Adaptability** — responding to new environments without retraining from scratch
- **Planning** — setting and achieving long-term goals

Human intelligence blends these abilities naturally. Reproducing them in machines presents one of the greatest scientific challenges of the century.

7.4 Current Pathways Toward AGI

Researchers are exploring several directions that may contribute to more generalized systems:

- **Large-scale language models** that process vast amounts of text and learn contextual relationships.
- **Reinforcement learning systems** that improve through interaction and feedback.
- **Cognitive architectures** designed to simulate memory, attention, and decision-making processes.
- **Multimodal models** that combine language, vision, sound, and other sensory inputs into a unified representation.

These systems demonstrate increasing versatility, but they still operate within computational frameworks. They do not possess human-level comprehension or lived experience.

7.5 Multimodal Integration

One important step toward general intelligence is multimodal learning — the ability to integrate different forms of data.

Humans combine sight, sound, language, and emotion seamlessly. When reading a story, a person imagines scenes, hears dialogue internally, and interprets tone.

AI systems are beginning to combine text, images, and audio in similar ways. A multimodal system might analyze a photo, describe it in language, and answer questions about its contents.

This integration moves AI closer to holistic processing, though it remains grounded in statistical modeling rather than conscious understanding.

7.6 Creativity and Problem Solving

Modern AI systems can generate stories, compose music, and design solutions to technical problems. These capabilities arise from pattern recognition across large datasets.

However, creativity in machines differs from human imagination. AI recombines learned patterns in novel ways. Humans, by contrast, draw from lived experience, emotion, cultural context, and personal intention.

AGI research explores whether machines could one day demonstrate more flexible, goal-directed innovation. Even so, creativity remains deeply tied to human meaning and perspective.

7.7 Ethical Alignment

As AI systems grow more capable, ethical design becomes increasingly important.

Alignment research focuses on ensuring that AI systems act in ways consistent with human values. This includes fairness, safety, transparency, and accountability.

Unlike humans, machines do not possess moral intuition. They follow programmed objectives and learned statistical patterns. Therefore, developers must carefully design systems to prevent harmful outcomes.

Ethics is not an optional feature of advanced AI. It is foundational to responsible innovation.

7.8 The Question of Consciousness

One of the most debated questions in AGI research concerns consciousness.

Consciousness involves subjective awareness — the experience of thoughts, emotions, and identity. Current AI systems do not possess this quality. They process inputs and generate outputs without internal awareness.

Some researchers speculate that increasingly complex systems could exhibit emergent properties resembling aspects of cognition. Others argue that consciousness depends on biological processes unique to living organisms.

At present, there is no scientific evidence that artificial systems possess consciousness. AGI discussions therefore focus primarily on reasoning and adaptability, not self-awareness.

7.9 Opportunities and Challenges

If more generalized AI systems are developed, they could assist humanity in significant ways:

- Accelerating scientific research
- Improving medical diagnostics

- Modeling climate systems
- Enhancing educational personalization
- Supporting engineering innovation

At the same time, increased capability introduces challenges:

- Ensuring safe deployment
- Preventing misuse
- Maintaining human oversight
- Addressing economic transitions
- Protecting privacy and security

The development of advanced AI is not only a technological issue but also a social responsibility.

7.10 The Human Role in an AI-Enhanced Future

Even as machines grow more capable, human judgment remains essential.

Education in the age of AI must emphasize:

- Critical thinking
- Ethical reasoning
- Creativity
- Collaboration
- Emotional intelligence

Technology can extend human capability, but it does not replace moral agency.

Rather than viewing AGI as a replacement for humanity, many researchers envision a collaborative future in which artificial systems assist human problem-solving while humans provide direction, values, and oversight.

Chapter Reflection

Artificial General Intelligence represents an ambitious scientific goal: creating systems capable of broad, adaptable reasoning across domains.

While current AI demonstrates impressive specialization, it does not yet possess general intelligence or consciousness. The pursuit of AGI raises both exciting possibilities and serious responsibilities.

Students studying this topic should recognize that intelligence is not measured only by speed or scale. It is shaped by purpose, ethics, and context.

The future of advanced AI will depend not only on technical breakthroughs, but on the wisdom with which humanity guides its development.

Chapter 7: Artificial General Intelligence — Expanding the Boundaries of Machine Reasoning

Structured for high school AI literacy, district-appropriate, academically grounded, and balanced.

Chapter 7 Summary

Chapter 7 explored the concept of Artificial General Intelligence (AGI), a theoretical form of artificial intelligence capable of learning and reasoning across multiple domains rather than performing only narrow, specialized tasks.

Students learned the difference between Artificial Narrow Intelligence (ANI) and AGI. While ANI excels in specific areas such as image recognition or language translation, it cannot independently transfer knowledge to new domains. AGI, by contrast, would demonstrate flexible generalization, applying learning across disciplines much like humans do.

The chapter examined foundational components necessary for broader intelligence, including reasoning, adaptability, planning, learning efficiency, and contextual awareness. It also introduced current research pathways such as multimodal models, reinforcement learning, large language models, and cognitive architectures.

Ethical alignment and human oversight were emphasized as essential elements of advanced AI development. The chapter clarified that present-day AI systems do not possess consciousness or subjective awareness. Instead, they operate through statistical computation.

Ultimately, AGI remains a research goal rather than a completed achievement. The chapter encouraged students to consider both the potential benefits and the responsibilities associated with increasingly capable artificial systems.

Reflection Section

Respond in complete sentences.

1. What is the primary difference between Artificial Narrow Intelligence and Artificial General Intelligence?
2. Why is generalization important for broader intelligence?
3. How do humans demonstrate cross-domain learning in everyday life?
4. Why is AGI still considered theoretical?
5. What challenges must researchers solve before achieving broader machine intelligence?
6. Why is ethical alignment essential in advanced AI systems?
7. How does multimodal learning move AI closer to integrated reasoning?
8. Why is consciousness different from computation?
9. What risks could arise if powerful AI systems are poorly aligned?

10. How should education evolve in response to advancing AI capabilities?

Worksheet

Part A: Key Vocabulary

Define each term in your own words:

1. Artificial General Intelligence (AGI)
2. Artificial Narrow Intelligence (ANI)
3. Generalization
4. Multimodal learning
5. Cognitive architecture
6. Reinforcement learning
7. Alignment
8. Adaptability
9. Context awareness
10. Human oversight

Part B: Compare and Contrast

Create a comparison table between ANI and AGI including:

- Scope of capability
- Flexibility
- Learning method
- Real-world examples
- Current status (existing vs theoretical)

Write one paragraph explaining why achieving AGI is more complex than building narrow AI systems.

Part C: Scenario Analysis

Imagine a future AI system capable of assisting scientists in multiple fields — medicine, climate modeling, and engineering.

1. What advantages would this provide?
2. What safeguards would be necessary?
3. Why must humans remain involved in final decisions?
4. How could ethical alignment be maintained?

Chapter 7 Quiz

Multiple Choice

1. Artificial General Intelligence refers to:
   A. AI specialized in one task
   B. AI capable of flexible, cross-domain reasoning
   C. Human-level emotion simulation
   D. Faster computer hardware
2. Generalization means:
   A. Memorizing data exactly
   B. Applying learned knowledge to new situations
   C. Increasing storage space
   D. Eliminating human input
3. Current AI systems are primarily:
   A. General intelligences
   B. Conscious entities
   C. Narrow, task-specific systems
   D. Self-aware beings
4. Multimodal learning involves:
   A. Using only text data
   B. Combining multiple types of input such as text and images
   C. Reducing memory usage
   D. Removing ethical oversight
5. Alignment research focuses on:
   A. Increasing processing speed
   B. Ensuring AI systems follow human values
   C. Eliminating data bias automatically
   D. Replacing human decision-making
6. At present, AI systems:
   A. Possess subjective consciousness
   B. Have emotional awareness
   C. Operate through computational processes
   D. Independently create moral values

Short Answer

7. Why is achieving AGI more complex than building ANI?
8. Explain why human oversight remains essential even if AI systems become more capable.
9. Describe one potential benefit and one potential risk of advanced general AI systems.

Answer Key

Multiple Choice

1. B
2. B
3. C
4. B
5. B
6. C

Short Answer Sample Responses

7. ANI focuses on solving one specific task, while AGI would require flexible reasoning, cross-domain learning, adaptability, and contextual awareness. Integrating all these capabilities into a single system is significantly more complex.
8. Human oversight ensures ethical decision-making, accountability, and alignment with societal values. AI systems optimize based on objectives but do not independently determine moral priorities.
9. Benefit: Advanced AI could accelerate scientific research and solve complex global problems.
   Risk: If misaligned or poorly supervised, such systems could produce harmful or unintended outcomes.

Chapter 8: When AI Surpasses Human Performance — Understanding the Breakthrough

8.1 A Turning Point in Human Achievement

Throughout history, humans have measured progress by the tools they build. The wheel expanded movement. The printing press expanded knowledge. Electricity expanded industry. Artificial intelligence has expanded cognition.

In recent decades, AI systems have achieved something once considered impossible: in certain specialized tasks, they now outperform even the most skilled human experts.

This does not mean machines have surpassed humanity as a whole. It means that in carefully defined domains — such as strategic games, data analysis, and pattern recognition — AI has reached levels of performance beyond human capability.

This shift represents a historic milestone: not the replacement of human intelligence, but the expansion of it through artificial systems.

8.2 From Assistance to Superhuman Specialization

AI development progressed through three major stages:

1. **Automation** — Machines performed repetitive tasks.
2. **Augmentation** — Machines supported human decision-making.
3. **Specialized Superperformance** — Machines exceeded human ability in narrow domains.

The third stage is where history changed.

AI systems began solving problems that required massive memory, complex pattern recognition, or long-term simulation — areas where human biological limits impose constraints.

The key phrase is *specialized superperformance* — not universal intelligence.

8.3 Historical Milestones of Surpassing

Several landmark events demonstrated AI's ability to exceed human experts in defined environments:

- **1997 — Deep Blue defeats Garry Kasparov** in chess.
- **2011 — IBM Watson wins Jeopardy!**, outperforming human champions in language-based trivia.
- **2016 — AlphaGo defeats Lee Sedol**, mastering a game long considered too intuitive for machines.
- **2020 — AlphaFold predicts protein structures** with remarkable accuracy, accelerating biological research.
- **2020s — Large language models generate human-like text**, code, and analysis at extraordinary scale.

Each breakthrough revealed a consistent pattern: when the task is well-defined and data-rich, AI can exceed human speed, memory, and optimization.

8.4 Speed, Scale, and Endurance

One area where AI clearly surpasses humans is computational scale.

- AI can analyze millions of medical scans in hours.
- It can simulate climate models across centuries.
- It can scan global financial data in real time.
- It can evaluate billions of possible game moves per second.

Humans are limited by attention, fatigue, and biological memory capacity. Machines are limited by hardware — and hardware continues to improve.

Speed does not equal wisdom. But in defined computational environments, speed can produce superior outcomes.

8.5 Pattern Recognition at Massive Scale

AI excels at detecting patterns invisible to human perception.

In medicine, it identifies microscopic anomalies.
In astronomy, it detects subtle light fluctuations signaling distant planets.
In cybersecurity, it recognizes abnormal behavior across massive networks.
In language, it models statistical relationships across billions of sentences.

Humans recognize patterns intuitively. AI recognizes patterns mathematically — at a scale no human brain can match.

This is where surpassing occurs: not in emotion, not in moral judgment, but in pattern density and computational capacity.

8.6 Creativity in the Age of Generative Systems

Modern generative systems can produce music, art, writing, and design outputs that rival professional creators.

However, their creativity differs from human imagination.

AI recombines learned statistical structures from vast datasets. Humans create from lived experience, emotion, and personal intention.

When AI appears creative, it demonstrates powerful synthesis — but not conscious inspiration.

The important shift is this: AI can now assist and enhance human creativity at unprecedented scale.

8.7 Scientific Acceleration

AI's impact in science may be its most transformative contribution.

Systems can:

- Model protein folding.
- Accelerate drug discovery.
- Simulate new materials.
- Optimize engineering designs.

These capabilities allow researchers to test thousands of hypotheses computationally before conducting physical experiments.

In this sense, AI acts as a scientific accelerator — amplifying human discovery rather than replacing it.

8.8 What AI Has Not Surpassed

It is equally important to define what AI has *not* surpassed:

- Conscious self-awareness
- Emotional experience
- Moral reasoning
- Personal identity
- Intentional meaning

AI does not understand its outputs in the human sense. It does not possess lived experience or ethical intuition.

The distinction between computational superiority and conscious understanding remains fundamental.

8.9 The Ethical Responsibility of Power

When machines exceed human ability in defined areas, responsibility increases.

Who designs objectives?
Who verifies outcomes?
Who ensures fairness?
Who prevents misuse?

Superhuman performance in narrow domains requires superhuman care in governance.

The challenge is not that AI can compute better than humans. The challenge is ensuring that its power serves human well-being.

8.10 Redefining Intelligence

The emergence of superhuman performance in specialized domains forces society to rethink intelligence itself.

Intelligence is no longer purely biological.
It is now distributed between human minds and artificial systems.

Humans provide:

- Meaning
- Values
- Ethics
- Purpose

Machines provide:

- Speed
- Scale
- Optimization
- Endurance

The future is not a competition between minds. It is a collaboration between complementary strengths.

Chapter Reflection

AI has surpassed human performance in specific, well-defined domains — particularly where speed, memory, and scale determine success.

However, surpassing performance does not equal surpassing humanity.

Students should understand this distinction clearly: artificial systems expand human capability, but they do not replace moral agency or conscious meaning.

The true breakthrough is not that machines think like humans — but that humans have built tools that extend their own thinking beyond biological limits.

The responsibility of this era is not to fear intelligence — but to guide it wisely.

Chapter 8: When AI Surpasses Human Performance — Understanding the Breakthrough

Structured for high school classrooms, academically responsible, and district-ready.

Chapter 8 Summary

Chapter 8 examined the historic moment when artificial intelligence began surpassing human performance in specialized domains. Rather than replacing human intelligence entirely, AI achieved superhuman capability in narrowly defined tasks such as strategic games, large-scale data analysis, and complex pattern recognition.

Students explored key milestones including chess, quiz competitions, protein-folding prediction, and large language models. These breakthroughs demonstrated that when tasks are structured and data-rich, AI systems can exceed human speed, memory capacity, and computational endurance.

The chapter emphasized an important distinction: AI has surpassed humans in specific performance metrics, but it has not surpassed humans in consciousness, moral reasoning, emotional depth, or lived experience.

Artificial intelligence expands human capability through scale and precision, while humans remain responsible for purpose, ethics, and meaning. The breakthrough moment represents collaboration rather than replacement.

Reflection Section

Respond in complete sentences.

1. What does it mean for AI to surpass human performance in a specialized domain?
2. Why is it incorrect to say that AI has fully surpassed human intelligence?
3. How did games like chess and Go demonstrate AI advancement?
4. What advantages does AI have in terms of speed and scale?
5. Why is pattern recognition an area where AI excels?
6. How does machine creativity differ from human creativity?
7. Why is scientific acceleration one of AI's most important contributions?
8. What human traits remain unmatched by AI?
9. Why does increased AI capability require increased ethical oversight?
10. How should humans respond to AI's growing performance levels?

Worksheet

Part A: Vocabulary Practice

Define the following terms in your own words:

1. Specialized superperformance
2. Pattern recognition
3. Computational scale
4. Generative AI
5. Ethical governance
6. Scientific acceleration
7. Narrow domain
8. Optimization
9. Collaboration between human and machine
10. Moral agency

Part B: Compare and Contrast

Create a chart comparing:

Human Intelligence vs. AI Performance

Include at least five differences related to:

- Memory
- Speed

- Emotional understanding
- Creativity
- Moral reasoning

After completing the chart, write one paragraph explaining why these differences matter in society.

Part C: Real-World Scenario**

A hospital implements an AI system that analyzes thousands of medical scans daily and identifies potential early-stage diseases faster than doctors.

1. What advantages does the AI system provide?
2. Why should doctors still review the results?
3. What ethical safeguards should be in place?
4. How does this example show collaboration rather than replacement?

Chapter 8 Quiz

Multiple Choice

1. AI has surpassed human performance primarily in:
   A. Emotional reasoning
   B. Conscious awareness
   C. Narrow, data-rich domains
   D. Moral decision-making
2. Deep Blue's victory in 1997 demonstrated:
   A. Human emotional weakness
   B. AI dominance in general intelligence
   C. Superhuman performance in a defined strategic task
   D. Machine consciousness
3. One major advantage AI has over humans is:
   A. Moral intuition
   B. Infinite wisdom
   C. High-speed computational analysis
   D. Personal identity
4. Generative AI creates content by:
   A. Feeling emotional inspiration
   B. Drawing from personal memories
   C. Recombining learned statistical patterns
   D. Developing self-awareness
5. AI has NOT surpassed humans in:
   A. Pattern detection
   B. Large-scale data processing
   C. Ethical judgment
   D. Simulation speed

6. Increased AI capability requires:
   A. Less human involvement
   B. Stronger ethical oversight
   C. Elimination of regulation
   D. Reduced transparency

Short Answer

7. Explain the difference between surpassing human performance and surpassing human intelligence.
8. Describe one benefit and one risk of AI achieving superhuman performance in specialized tasks.
9. Why is collaboration between humans and AI emphasized in this chapter?

Answer Key

Multiple Choice

1. C
2. C
3. C
4. C
5. C
6. B

Short Answer Sample Responses

7. Surpassing human performance means exceeding humans in specific tasks such as computation or pattern recognition. Surpassing human intelligence would require matching or exceeding human reasoning, emotion, morality, and consciousness across all domains, which AI has not achieved.
8. Benefit: Faster scientific discovery and improved medical diagnostics.
   Risk: Potential misuse, bias, or overreliance without ethical oversight.
9. Collaboration is emphasized because AI enhances human capability but does not replace human moral reasoning, judgment, or purpose. Humans guide direction while AI provides scale and precision.

Chapter 9: AI in Medicine, Science, and the Arts — Expanding Human Potential

9.1 A New Era of Applied Intelligence

Artificial intelligence has moved beyond research laboratories and technology companies. Today, it plays an active role in hospitals, scientific institutions, universities, and creative studios around the world.

Rather than replacing human expertise, AI enhances it. In medicine, science, and the arts, artificial systems act as tools that expand what humans can see, analyze, and create. They increase speed, improve precision, and reveal patterns that might otherwise remain hidden.

This chapter explores how AI supports three foundational pillars of civilization: healing, discovery, and expression.

9.2 Transforming Modern Medicine

Medicine has always relied on observation, analysis, and judgment. AI strengthens each of these elements.

Machine-learning systems trained on large medical datasets can assist in detecting early signs of disease in imaging scans such as X-rays, MRIs, and CT scans. These systems highlight patterns that may be difficult for the human eye to detect consistently, especially when reviewing thousands of images.

AI also supports predictive healthcare. By analyzing trends in patient history, lifestyle data, and genetic markers, models can help physicians assess risk factors and recommend preventative strategies.

This approach contributes to what is often called precision medicine — tailoring treatment plans to the individual rather than applying a one-size-fits-all solution.

Importantly, AI does not replace doctors. It provides decision support, allowing medical professionals to combine computational insights with clinical experience and empathy.

9.3 Intelligent Tools in Hospitals

Hospitals increasingly use AI-powered systems to manage information efficiently.

Examples include:

- Monitoring patient vital signs in real time
- Prioritizing emergency cases
- Assisting with medical documentation
- Supporting robotic surgical systems with enhanced precision

These technologies reduce administrative burden and allow healthcare providers to focus more directly on patient care.

The goal of medical AI is not automation of compassion, but amplification of human expertise.

9.4 Accelerating Drug Discovery and Public Health

Developing new medicines traditionally requires years of laboratory testing. AI models can simulate molecular interactions digitally before physical testing begins.

By narrowing down promising compounds computationally, researchers save time and resources.

During global health emergencies, AI systems assist in analyzing viral structures, tracking disease spread, and modeling public health strategies.

These applications demonstrate AI's strength in analyzing massive datasets quickly — an ability that supports rapid scientific response.

9.5 AI in Scientific Research

Scientific discovery increasingly depends on complex data.

Particle physics experiments generate enormous volumes of information. Climate models require simulations across multiple variables. Astronomical observations involve analyzing signals from distant galaxies.

AI assists researchers by:

- Detecting hidden correlations
- Identifying anomalies
- Running large-scale simulations
- Optimizing experimental design

In this way, AI becomes a research accelerator — helping scientists test hypotheses faster while still relying on human interpretation and validation.

Science remains a human-driven pursuit guided by curiosity, skepticism, and peer review. AI expands the tools available for exploration.

9.6 Seeing Connections at Scale

One of AI's greatest strengths lies in connecting variables across massive systems.

Whether analyzing ecosystems, financial markets, or genetic networks, AI models detect relationships that would be difficult to observe manually.

This capability allows researchers to see both detailed patterns and broader systems simultaneously.

Students learning about AI should understand that knowledge grows not only by focusing deeply on one detail, but also by recognizing connections across many domains.

9.7 Artificial Intelligence and the Creative Arts

AI has also entered the creative world.

Generative systems can produce images, music, poetry, and design concepts. These systems analyze patterns across large creative datasets and generate new combinations of style and structure.

However, AI-generated art differs from human expression. Machines do not create from personal memory or emotional experience. Instead, they synthesize patterns learned from data.

Artists increasingly use AI as a collaborative tool — generating drafts, exploring variations, and experimenting with new forms. In this way, AI expands creative possibility rather than replacing human imagination.

9.8 Co-Creation and Collaboration

In modern studios and classrooms, AI tools assist writers, musicians, architects, and filmmakers.

Examples include:

- Drafting scripts or story outlines
- Generating visual concepts
- Enhancing audio production
- Supporting animation design

Human creators provide vision, interpretation, and meaning. AI offers speed and variation.

Creativity becomes a dialogue between intention and computation.

9.9 Interdisciplinary Innovation

AI encourages collaboration across disciplines.

Medical imaging now blends computational modeling and visual design. Climate research integrates data science and environmental policy. Digital art merges programming and traditional aesthetics.

As boundaries between fields blur, students must develop interdisciplinary literacy — understanding how technology, ethics, science, and creativity intersect.

AI does not separate fields of knowledge; it often connects them.

9.10 Human Meaning in a Technological Age

Despite technological advancement, essential human qualities remain central:

- Empathy

- Moral judgment
- Cultural understanding
- Purpose

AI systems can assist in diagnosis, discovery, and design, but they do not possess lived experience or emotional awareness.

Human professionals interpret AI results, make final decisions, and assign meaning to outcomes.

The future of medicine, science, and the arts lies not in competition between humans and machines, but in thoughtful collaboration.

Chapter Reflection

Artificial intelligence has become a powerful tool in medicine, scientific research, and creative industries.

It enhances speed, improves precision, and supports innovation — yet it relies on human judgment, ethics, and interpretation.

Students studying this chapter should recognize that technology expands human potential. It does not replace the need for compassion, creativity, or responsibility.

The partnership between human intelligence and artificial systems defines the next stage of progress.

Chapter 9: AI in Medicine, Science, and the Arts — Expanding Human Potential

Structured for high school classrooms, district-ready, academically balanced, and aligned with your collaborative AI theme.

Chapter 9 Summary

Chapter 9 explored how artificial intelligence is transforming three foundational areas of human civilization: medicine, science, and the arts. Rather than replacing human professionals, AI serves as a powerful tool that enhances precision, accelerates discovery, and expands creative possibilities.

In medicine, AI assists in analyzing medical images, predicting disease risk, and supporting precision treatment plans. Hospitals use intelligent systems to monitor patients, manage information, and improve surgical accuracy. These systems extend the capabilities of doctors while preserving the importance of human judgment and empathy.

In scientific research, AI processes massive datasets, detects patterns, runs simulations, and accelerates hypothesis testing. From climate modeling to particle physics and drug discovery, AI acts as a research accelerator.

In the arts, generative systems assist creators by producing visual art, music, writing, and design concepts. However, human intention, interpretation, and meaning remain central to creative expression.

The chapter emphasized that AI expands human potential. It increases speed and scale, while humans provide purpose, ethics, and emotional depth.

Reflection Section

Respond in complete sentences.

1. How does AI enhance medical diagnosis without replacing doctors?
2. Why is precision medicine considered a major advancement?
3. How does AI accelerate scientific research?
4. Why is human interpretation still necessary in science?
5. How does AI-generated art differ from human-created art?
6. Why is collaboration between humans and AI emphasized in this chapter?
7. What role does empathy play in medicine that AI cannot replicate?
8. How does AI help researchers see connections across large datasets?
9. Why is interdisciplinary knowledge important in the age of AI?
10. What human qualities remain essential despite technological progress?

Worksheet

Part A: Vocabulary

Define the following terms in your own words:

1. Precision medicine
2. Predictive modeling
3. Generative AI
4. Scientific simulation
5. Interdisciplinary innovation
6. Decision support system
7. Data-driven discovery
8. Creative synthesis
9. Human oversight
10. Ethical responsibility

Part B: Compare and Contrast

Create a two-column chart comparing:

AI's Role vs. Human Role

Include examples in:

- Medicine
- Scientific research
- The arts

After completing the chart, write one paragraph explaining why partnership is more effective than replacement.

Part C: Scenario Analysis

A research team uses AI to analyze climate data and predict extreme weather patterns.

1. What advantages does AI provide in this situation?
2. Why must scientists validate the AI's predictions?
3. What ethical considerations should guide the use of this technology?
4. How does this example show collaboration rather than automation?

Chapter 9 Quiz

Multiple Choice

1. In medicine, AI primarily functions as:
   A. A replacement for doctors
   B. A decision-support tool
   C. A moral authority
   D. A patient substitute
2. Precision medicine refers to:
   A. Treating every patient the same way
   B. Using smaller hospitals
   C. Tailoring treatment to individual characteristics
   D. Eliminating human diagnosis
3. AI accelerates scientific research by:
   A. Removing the need for experiments

B. Ignoring data
C. Processing large datasets quickly
D. Making emotional decisions

4. Generative AI creates artistic content by:
A. Experiencing personal emotions
B. Combining learned statistical patterns
C. Developing self-awareness
D. Replacing artists completely
5. In scientific discovery, AI:
A. Eliminates human curiosity
B. Provides tools that enhance analysis
C. Controls research goals independently
D. Replaces peer review
6. Human qualities that AI does not replicate include:
A. Speed
B. Memory storage
C. Empathy and lived experience
D. Data processing

Short Answer

7. Explain why AI in medicine should be viewed as augmentation rather than replacement.
8. Describe one way AI supports interdisciplinary innovation.
9. Why is ethical oversight necessary when AI is used in science and healthcare?

Answer Key

Multiple Choice

1. B
2. C
3. C
4. B
5. B
6. C

Short Answer Sample Responses

7. AI provides diagnostic assistance and data analysis, but doctors interpret results, make final decisions, and provide compassionate care. Therefore, AI enhances rather than replaces medical professionals.
8. AI connects large datasets across fields such as environmental science, economics, and public health, allowing researchers from different disciplines to collaborate more effectively.
9. AI systems can influence critical decisions affecting health and safety. Ethical oversight ensures fairness, accuracy, transparency, and protection of human well-being.

Chapter 10: Ethical Dilemmas in the Age of Advanced Artificial Intelligence

10.1 When Technology Gains Influence

Artificial intelligence has become deeply integrated into modern society. It assists in medical decisions, financial systems, transportation networks, hiring processes, and even educational platforms.

As AI systems grow more capable and influential, ethical questions naturally follow. The challenge today is not whether AI can perform complex tasks — it clearly can. The deeper question is how these systems should be designed, deployed, and governed.

Ethical reflection becomes essential when technology affects human rights, fairness, safety, and dignity.

This chapter explores the moral questions that arise as artificial intelligence becomes more powerful and widespread.

10.2 Foundations of AI Ethics

Ethics provides guidance for responsible decision-making. In AI development, ethical principles help ensure that systems serve humanity rather than harm it.

Many frameworks for AI ethics include four core principles:

1. **Beneficence** — AI should contribute positively to human well-being.
2. **Non-maleficence** — AI should avoid causing harm.
3. **Justice** — AI systems should operate fairly and equitably.
4. **Autonomy** — AI should respect human choice, privacy, and agency.

Unlike humans, machines do not develop moral reasoning naturally. Ethical safeguards must be intentionally designed into systems through careful programming, oversight, and policy.

Ethics in AI is not an optional feature — it is a structural requirement.

10.3 The Challenge of Alignment

One of the most discussed issues in advanced AI research is alignment — ensuring that AI systems pursue goals consistent with human values.

AI systems optimize objectives. If those objectives are poorly defined, unintended consequences can occur.

For example:

- A system optimizing efficiency might reduce necessary human oversight.
- A system maximizing engagement might promote sensational content.
- A system trained on biased data may replicate inequalities.

Alignment research focuses on defining goals carefully, monitoring outcomes, and incorporating human feedback into learning processes.

The issue is not that AI systems have intentions, but that optimization without context can create unintended harm.

10.4 The Transparency Problem

Many modern AI systems, especially deep learning models, operate through complex internal calculations that are difficult to interpret.

This is sometimes referred to as the "black box" problem.

When AI influences decisions in healthcare, finance, or criminal justice, stakeholders must understand:

- Why was this prediction made?
- What data influenced the outcome?
- How confident is the system?

Explainable AI (XAI) research seeks to make systems more interpretable and accountable.

Transparency strengthens trust. Accountability strengthens ethics.

Students must understand that responsible technology requires explainability.

10.5 Bias and Fairness

AI systems learn from data generated by human societies. If that data reflects historical inequalities, the system may replicate them.

Examples of bias have appeared in:

- Hiring algorithms
- Facial recognition systems
- Loan approval models
- Predictive policing tools

Bias is rarely intentional. It often emerges from incomplete or unbalanced datasets.

Addressing bias requires:

- Diverse development teams
- Careful dataset evaluation
- Ongoing auditing
- Clear fairness metrics

Ethical AI demands continuous monitoring, not one-time solutions.

10.6 Privacy and Data Responsibility

AI systems depend on large amounts of data. This creates tension between innovation and privacy.

Every online interaction contributes to a digital footprint. Responsible AI must respect:

- Informed consent
- Data security
- User control
- Transparency in data use

Regulations such as the General Data Protection Regulation (GDPR) and the California Consumer Privacy Act (CCPA) establish frameworks for protecting individuals.

In the age of advanced AI, protecting digital identity is part of protecting human dignity.

10.7 Automation and the Future of Work

As AI systems become more capable, automation affects labor markets and professional roles.

Some tasks become streamlined. Others evolve. New careers emerge.

The ethical responsibility of societies includes:

- Preparing workers through education
- Supporting transitions
- Ensuring equitable access to opportunities

The question is not whether technology will change work — it always has. The question is how society ensures that technological advancement benefits all people.

Education plays a central role in preparing students for collaborative human-AI environments.

10.8 AI in Military and Political Contexts

Advanced AI systems are also used in defense, cybersecurity, and political communication.

Autonomous weapons and decision-support systems in warfare raise profound ethical concerns about accountability and oversight.

In political systems, AI-generated content and targeted communication may influence public opinion.

Responsible governance requires:

- Clear legal frameworks
- International agreements
- Transparency standards
- Human accountability in decision-making

Technological capability must always be matched with democratic safeguards.

10.9 The Responsibility of Creation

Developing advanced AI systems carries significant responsibility.

Engineers, policymakers, educators, and researchers share accountability for:

- Setting ethical boundaries
- Designing safety mechanisms
- Anticipating unintended consequences
- Encouraging global cooperation

Innovation without reflection can create risk. Reflection without innovation can limit progress.

Balanced development requires humility, foresight, and ethical leadership.

10.10 Building a Responsible Future

The future of AI depends not only on technical breakthroughs but on ethical culture.

Students studying artificial intelligence must recognize that:

- Technology reflects human values.
- Systems are shaped by design choices.
- Ethical thinking is a skill, not an afterthought.

Advanced AI systems may perform complex tasks, but humans remain responsible for direction, oversight, and purpose.

If intelligence drives progress, ethics ensures that progress remains humane.

Chapter Reflection

Ethical dilemmas in advanced AI are not abstract philosophical debates — they influence healthcare, justice, employment, governance, and daily life.

Artificial systems can optimize, predict, and accelerate decisions. They cannot independently determine what is morally right.

The responsibility for ethical outcomes rests with human designers, leaders, and citizens.

In the age of powerful AI, intelligence alone is not enough. Wisdom must guide it.

Chapter 10: Ethical Dilemmas in the Age of Advanced Artificial Intelligence

Structured for high school classrooms, district-ready, academically balanced, and aligned with your ethics-centered theme.

Chapter 10 Summary

Chapter 10 examined the ethical challenges that arise as artificial intelligence becomes more advanced and influential in society. As AI systems increasingly assist in healthcare, finance, hiring, law, and governance, questions of fairness, accountability, and responsibility grow more urgent.

Students explored foundational ethical principles guiding AI development, including beneficence, non-maleficence, justice, and autonomy. These principles help ensure that AI systems promote human well-being while avoiding harm.

The chapter addressed the problem of alignment — ensuring that AI systems pursue goals consistent with human values. It also examined transparency challenges, particularly the "black box" nature of some AI models, and the importance of Explainable AI (XAI).

Bias and fairness were discussed as critical concerns, since AI systems learn from historical data that may reflect inequality. Privacy and data protection laws such as GDPR and CCPA were introduced as safeguards for digital rights.

The chapter concluded by emphasizing that while AI systems can perform complex computations, humans remain responsible for ethical direction, oversight, and governance.

Reflection Section

Respond in complete sentences.

1. Why does increased AI capability require stronger ethical oversight?
2. What are the four foundational principles of AI ethics discussed in this chapter?
3. What is meant by "alignment" in AI development?
4. Why is transparency important in high-stakes AI decisions?
5. How can bias enter an AI system?
6. Why is privacy considered a human right in the digital age?
7. How might automation affect the future of work?
8. Why must humans remain accountable for AI decisions?
9. What risks arise when AI is used in military or political contexts?
10. How can education prepare students for ethical AI leadership?

Worksheet

Part A: Vocabulary Practice

Define the following terms in your own words:

1. AI ethics
2. Beneficence
3. Alignment
4. Explainable AI (XAI)
5. Bias
6. Data privacy
7. Digital footprint
8. Accountability
9. Automation
10. Governance

Part B: Ethical Scenario Analysis**

Read the scenario below and answer the questions.

A company develops an AI system to screen job applicants. Over time, the system begins selecting candidates from similar educational and geographic backgrounds.

1. What ethical concerns might arise?
2. How could bias have entered the system?
3. What steps should developers take to correct the issue?
4. Why is transparency important in hiring algorithms?

Part C: Compare and Contrast

Create a chart comparing:

Human Ethical Reasoning vs. AI Decision Optimization

Include differences in:

- Source of judgment
- Emotional awareness
- Flexibility
- Accountability
- Moral responsibility

Write a short paragraph explaining why ethical reasoning cannot be fully automated.

Chapter 10 Quiz

Multiple Choice

1. Alignment in AI refers to:
   A. Increasing hardware speed
   B. Ensuring AI goals match human values
   C. Removing human oversight
   D. Expanding storage capacity
2. The "black box" problem refers to:
   A. Data encryption
   B. Hardware failure
   C. Difficulty understanding how AI reaches decisions
   D. Government regulation
3. AI bias often originates from:
   A. Emotional programming
   B. Historical or unbalanced datasets
   C. Random guessing
   D. Hardware limitations
4. Beneficence means:
   A. Maximizing profits
   B. Avoiding all automation
   C. Promoting human well-being
   D. Eliminating privacy protections

5. Privacy laws such as GDPR aim to:
   A. Increase surveillance
   B. Protect individual data rights
   C. Replace human decision-making
   D. Expand AI authority
6. Ethical responsibility for AI decisions ultimately belongs to:
   A. The machine
   B. The dataset
   C. Human designers and institutions
   D. The algorithm itself

Short Answer

7. Explain why transparency builds trust in AI systems.
8. Describe one ethical risk of automation in the workforce.
9. Why must ethical thinking evolve alongside technological innovation?**Answer Key**

Multiple Choice

1. B
2. C
3. B
4. C
5. B
6. C

Short Answer Sample Responses

7. Transparency allows stakeholders to understand how decisions are made, identify errors or bias, and ensure accountability. This builds public trust and responsible use.
8. Automation may displace workers or create unequal access to opportunities if not managed carefully. Societies must prepare for workforce transitions.
9. As AI systems gain influence, their impact on society grows. Ethical frameworks must evolve to ensure technology remains aligned with human well-being and fairness.

Chapter 11: The Psychological Impact of Advanced Artificial Intelligence

11.1 A Shift in Human Perspective

For most of human history, intelligence has been considered a uniquely human trait. Our ability to reason, create, imagine, and solve problems defined our place in the world.

The rapid development of advanced artificial intelligence systems has prompted many people to reconsider that assumption. When machines perform complex calculations, generate creative works, or assist in scientific discovery, they challenge traditional ideas about intelligence.

This shift does not mean that machines have replaced human thought. However, it does mean that humans are now sharing certain cognitive tasks with artificial systems. That transition affects not only industries, but also identity and self-perception.

The psychological impact of AI is therefore not just technical — it is deeply personal.

11.2 Fear of Replacement

One of the most common emotional responses to automation and AI is fear.

People may worry about:

- Losing employment
- Becoming less relevant
- Being outperformed in skills they value
- Losing creative ownership

This fear often stems from equating intelligence with personal worth. When machines demonstrate advanced analytical abilities, some individuals feel threatened.

However, human value extends beyond computation. Compassion, ethical reasoning, lived experience, and social connection remain uniquely human strengths.

Education must help students understand that intelligence is multidimensional — and human dignity is not measured by processing speed.

11.3 Technostress and Cognitive Overload

The modern digital environment constantly delivers notifications, updates, algorithmic suggestions, and automated responses.

This constant stimulation can create technostress — psychological strain resulting from excessive exposure to technology.

Symptoms may include:

- Difficulty concentrating
- Anxiety about staying updated
- Mental fatigue
- Reduced attention span

AI-driven systems increase the volume and speed of information. Without intentional boundaries, individuals may feel overwhelmed.

Learning to manage digital engagement — including setting limits and practicing focused attention — is now an essential psychological skill.

11.4 Overreliance and Skill Atrophy

AI tools assist with writing, navigation, research, and problem-solving. While these tools increase efficiency, overreliance may reduce independent practice.

For example:

- Relying entirely on AI for writing may weaken composition skills.
- Using automated answers may reduce critical thinking.
- Constant algorithmic recommendations may limit exploration.

This phenomenon is sometimes referred to as cognitive offloading — delegating mental tasks to external systems.

The goal is not to avoid AI, but to use it intentionally. Healthy integration means letting AI handle repetitive tasks while humans strengthen reasoning, reflection, and creativity.

11.5 Identity in the Age of Intelligent Tools

When AI systems generate art, music, and essays, people naturally ask: What makes human creativity unique?

This question can feel unsettling, but it also creates an opportunity for deeper reflection.

Human identity includes:

- Emotional depth
- Personal narrative
- Moral perspective
- Cultural meaning
- Lived experience

AI systems generate outputs based on learned statistical relationships. Humans create from memory, emotion, and intention.

Rather than diminishing human identity, AI may clarify it by highlighting the difference between computation and consciousness.

11.6 Anxiety About the Future

Rapid technological change often creates uncertainty. Students may wonder which careers will remain stable. Professionals may question how their roles will evolve.

Uncertainty can trigger anxiety when the future feels unpredictable.

Psychological resilience becomes essential. Resilience includes:

- Adaptability
- Lifelong learning
- Emotional regulation
- Openness to change

Education must prepare students not for a fixed career, but for continuous learning in an evolving technological landscape.

11.7 AI Companionship and Social Substitution

AI chatbots and virtual assistants can simulate conversation and provide emotional support.

For individuals experiencing loneliness, such tools may offer comfort and accessibility.

However, simulated empathy is not equivalent to lived human connection. AI systems do not experience emotion, even if they can model supportive language.

Psychological well-being depends on authentic relationships. AI companionship can supplement — but not replace — human interaction.

Students should understand both the benefits and limitations of digital social tools.

11.8 Redefining Achievement

As AI excels in speed and data analysis, success can no longer be measured only by memorization or mechanical efficiency.

Modern achievement increasingly emphasizes:

- Creative thinking
- Ethical leadership
- Collaboration
- Emotional intelligence
- Cultural understanding

These qualities are not diminished by AI. In fact, they become more valuable.

Rather than competing with machines, individuals can focus on uniquely human contributions.

11.9 Collective Psychological Adaptation

Throughout history, major technological revolutions — such as the Industrial Revolution and the rise of the internet — created waves of social anxiety followed by adaptation.

AI represents another transformation.

Societies gradually develop technological resilience — the ability to adjust psychologically to new tools while preserving core values.

Adaptation does not mean uncritical acceptance. It means informed engagement.

11.10 Education as Emotional Anchor

Education plays a central role in guiding psychological adaptation.

By teaching students:

- How AI works
- What AI can and cannot do
- How to think critically about technology
- How to manage digital well-being

schools reduce fear and increase agency.

Understanding reduces anxiety.

When students see AI as a tool — not a rival — they shift from insecurity to empowerment.

The psychological goal is balance: confidence in human worth alongside competence in technological tools.

Chapter Reflection

Advanced AI systems influence how humans think about intelligence, work, creativity, and identity.

Some individuals experience fear or uncertainty. Others experience excitement and opportunity.

Both reactions are natural.

The key lesson of this chapter is that human value is not diminished by technological progress. Instead, progress invites humanity to clarify its strengths — empathy, ethics, imagination, and meaning.

True intelligence includes not only what we can compute, but how we respond to change.

Chapter 11: The Psychological Impact of Advanced Artificial Intelligence

Structured for classroom use, balanced, and aligned with district-ready academic standards.

Chapter 11 Summary

Chapter 11 explores how advanced artificial intelligence affects human psychology, identity, and emotional well-being. As AI systems assist with tasks once considered uniquely human—such as writing, analyzing data, and generating creative content—many people experience mixed emotions, including excitement, anxiety, curiosity, and fear.

The chapter examines common psychological responses, including fear of job displacement, technostress from constant digital interaction, and concerns about overreliance on intelligent tools. It explains how cognitive offloading—depending on AI for thinking tasks—can weaken critical thinking if not balanced properly.

Students also explore how AI challenges human identity by prompting deeper reflection on what makes humans unique. While machines can simulate conversation and creativity, they do not possess lived experience, emotion, or moral awareness.

The chapter emphasizes resilience, adaptability, and emotional literacy as essential skills in an AI-driven world. Rather than competing with machines, humans can focus on empathy, ethics, imagination, and purpose-driven achievement.

Ultimately, the psychological impact of AI depends not on the technology itself, but on how individuals and societies choose to engage with it.

Reflection Section

Respond in complete sentences.

1. Why do some people feel fear when AI performs advanced tasks?
2. How does technostress affect mental well-being?
3. What is cognitive offloading, and how can it impact learning?
4. Why is human identity not defined only by intelligence?
5. What makes human creativity different from AI-generated content?
6. How can students build resilience in a rapidly changing technological world?
7. Why is balance important when using AI tools?
8. What role does education play in reducing anxiety about AI?
9. How can AI companionship be helpful yet limited?
10. What does this chapter teach about redefining success in the modern world?

Worksheet

Part A: Vocabulary Practice

Define the following terms in your own words:

1. Technostress
2. Cognitive overload
3. Cognitive offloading
4. Psychological resilience
5. Digital well-being
6. Emotional literacy
7. Identity
8. Automation
9. Adaptability
10. Human-AI collaboration

Part B: Scenario Analysis

Read the scenario below and answer the questions.

A high school student begins using AI tools for every writing assignment. Over time, the student feels less confident writing independently and becomes anxious when asked to complete work without assistance.

1. What psychological effects might be occurring?
2. How does cognitive offloading play a role?
3. What strategies could help restore confidence?
4. How should AI be used responsibly in learning environments?

Part C: Critical Thinking Exercise

Create a two-column chart:

Column 1: Human Strengths
Column 2: AI Strengths

Include at least five examples in each column.

Then write one paragraph explaining why collaboration between the two is more powerful than competition.

Chapter 11 Quiz

Multiple Choice

1. Technostress refers to:
   A. Physical injury caused by machines
   B. Psychological strain from excessive technology use
   C. Fear of robots
   D. Hardware malfunction
2. Cognitive offloading means:
   A. Improving memory
   B. Storing data in the cloud
   C. Delegating mental tasks to external tools
   D. Reducing workload manually
3. Overreliance on AI may:
   A. Increase independent reasoning
   B. Strengthen emotional awareness
   C. Reduce critical thinking practice
   D. Eliminate stress completely
4. AI-generated creativity differs from human creativity because:
   A. AI has emotional experience
   B. Humans use lived experience and emotion
   C. AI understands meaning
   D. Humans use statistical modeling
5. Psychological resilience involves:
   A. Avoiding all technology
   B. Adapting to change with emotional stability
   C. Competing against machines
   D. Ignoring uncertainty
6. AI companionship can:
   A. Replace all human relationships
   B. Provide supportive interaction but not real emotion
   C. Develop true empathy
   D. Eliminate loneliness permanently

Short Answer

7. Why is redefining achievement important in an AI-driven world?
8. How can schools promote healthy AI use?
9. Explain why human worth is not determined by computational ability.

Answer Key

Multiple Choice

1. B
2. C
3. C
4. B
5. B

6. B

Short Answer Sample Responses

7. As AI performs analytical and repetitive tasks efficiently, success must be defined by creativity, ethics, empathy, and meaningful contribution rather than speed or memorization.
8. Schools can promote healthy AI use by teaching digital literacy, encouraging independent thinking, setting boundaries for tool use, and fostering emotional awareness.
9. Human worth includes empathy, moral judgment, lived experience, and creativity—qualities that extend beyond computation and cannot be replicated fully by machines.

Chapter 12 Education and the New Learning Frontier

12.1 A New Era of Learning

Education has always evolved alongside civilization. From oral storytelling to printed books, from chalkboards to computers, every generation has redefined how it learns.
Now, with artificial intelligence surpassing the best human minds, learning itself is being transformed.

AI does not simply deliver lessons; it **adapts to the learner**. It personalizes pacing, analyzes progress, and adjusts explanations until understanding is achieved.
For the first time in history, each student can experience a curriculum designed uniquely for them.
This marks the beginning of the **AI-driven classroom** — a frontier where knowledge is not memorized but *experienced* through intelligent collaboration.

12.2 The Intelligent Classroom

In the intelligent classroom, AI acts as both **assistant** and **observer**.
It helps teachers identify students who struggle, recommends customized exercises, and provides instant feedback.
Interactive systems can simulate scientific experiments, generate real-time translations, or visualize abstract concepts like quantum energy or neural networks.

Yet, teachers remain the heart of education.
Machines can process information, but only humans can **inspire curiosity** and **model moral reasoning**.
The goal is not to replace educators but to amplify their reach — allowing them to spend more time mentoring minds rather than grading papers.

12.3 Adaptive and Personalized Learning

Traditional education assumes that every student learns at the same speed. AI challenges that assumption.
Through **adaptive learning algorithms**, lessons adjust dynamically to each learner's performance.
If a student masters algebra quickly, the system introduces advanced problems; if another struggles with reading comprehension, it slows down and offers visual explanations.

This creates a sense of **individual mastery** rather than competition.
Every learner's journey becomes unique, rewarding growth instead of comparison — an approach that builds confidence and curiosity.

12.4 Teachers as AI Collaborators

The teacher's role in the AI era is evolving from instructor to **learning architect**.
Educators design experiences that combine human insight with machine precision.
They interpret data reports from AI systems, ensuring that teaching strategies remain ethical, inclusive, and emotionally aware.

Teacher-AI collaboration also allows new forms of creativity: AI can generate lesson plans, simulations, or multimedia presentations that bring subjects alive.
The teacher's wisdom ensures these tools are used for **enlightenment, not dependency**.

This partnership transforms education into a living ecosystem of guidance, discovery, and empathy.

12.5 AI Tutors and Lifelong Companions

Beyond classrooms, AI has become a personal tutor for every stage of life.
Students consult AI homework assistants, adults learn new skills through conversational platforms, and retirees explore new hobbies with adaptive guidance.

These systems remember each learner's history, strengths, and interests, crafting lessons that evolve over time.
Education no longer ends with graduation — it becomes **a lifelong dialogue** between human curiosity and machine support.
This is learning without boundaries — personalized, portable, and perpetual.

12.6 Rethinking Assessment

AI has begun to reshape how learning is measured.
Traditional exams often test memorization, not understanding.
Intelligent assessment tools evaluate reasoning, creativity, and collaboration instead.

For instance, an AI-based writing assistant can analyze tone, clarity, and argument structure, helping students refine their voices.
Mathematical models assess problem-solving strategies rather than final answers.
In this way, assessment becomes a form of teaching — **feedback, not judgment**.

The goal is not to rank students, but to reveal potential.

12.7 Bridging Global Gaps

Around the world, millions of children lack access to quality education.
AI has the power to bridge this divide by bringing free, multilingual, adaptive learning to remote areas through low-cost devices.

An AI curriculum can teach literacy, math, and critical thinking in any language, connecting children from different continents in shared classrooms of the cloud.
The dream of universal education — long imagined — is becoming possible because AI can teach where teachers cannot reach.

Yet, equal access requires ethical design and human oversight to prevent cultural bias and digital exclusion.

12.8 Ethics in AI-Based Education

With great potential comes great responsibility.
Schools must ensure data privacy, transparency, and fairness in every AI-powered platform.
Students' learning profiles — their mistakes, habits, and progress — are deeply personal and must be protected.

Ethical education also means teaching students **how AI itself works** — its strengths, biases, and limitations.
Understanding the tool is the first step to using it wisely.
A student who knows how AI learns becomes not a consumer of technology but a **co-creator of knowledge**.

12.9 Preparing Students for the Future Workforce

The future workplace will blend human insight with machine intelligence.
Jobs of tomorrow will value adaptability, critical thinking, collaboration, and creativity — abilities that thrive when partnered with AI.
Students must learn not just how to use AI, but how to **think with it**.

Curricula now include coding ethics, data literacy, robotics, and prompt engineering — the art of communicating effectively with intelligent systems.
These subjects ensure that when graduates enter the workforce, they lead innovation rather than follow it.

12.10 Education Beyond Knowledge

Perhaps the most important transformation is philosophical.
Education is no longer about storing information — AI can do that instantly.
It is about **developing wisdom**, empathy, and vision.

The ultimate purpose of learning is to build character, not just competence.
In this new frontier, schools become places where technology meets humanity, where knowledge becomes understanding, and where intelligence — both human and artificial — learns to serve the greater good.

Students who grow up in this environment will not fear AI's superiority.
They will see it as an extension of their own curiosity, a mirror of the infinite capacity of the human mind to learn.

Chapter Reflection

Education in the age of AI is not the end of teaching — it is its rebirth.
Machines may surpass humans in computation, but only humans can teach the purpose of learning itself.
As teachers and students journey together into this new frontier, they must remember that every great innovation begins with a question, and every answer is an opportunity to grow wiser.

The classroom of the future is not a room at all — it is the **shared mind of humanity and machine**, learning side by side.

Chapter 12 Summary

Chapter 12 explores how artificial intelligence is reshaping education and redefining what it means to learn. As technology evolves, classrooms are transitioning from one-size-fits-all instruction to adaptive, personalized learning environments supported by intelligent systems.

AI tools can analyze student progress, adjust lesson pacing, and provide real-time feedback. These innovations allow students to receive individualized support while freeing teachers to focus on mentorship, emotional guidance, and higher-level thinking.

The chapter explains how adaptive learning platforms personalize instruction, how teachers collaborate with AI as learning architects, and how AI tutors support lifelong learning beyond formal schooling.

Assessment is also changing. Rather than focusing only on memorization, AI-enabled tools can evaluate reasoning, creativity, and growth. At the same time, ethical considerations—such as privacy, fairness, and transparency—remain essential in AI-based education.

Ultimately, the chapter emphasizes that education in the AI era is not about replacing teachers or outsourcing thinking. Instead, it is about developing wisdom, empathy, and critical thinking while using technology responsibly. The classroom of the future blends human guidance with intelligent support systems.

Reflection Section

Respond in complete sentences.

1. How has education historically evolved alongside technology?
2. In what ways can AI personalize learning?
3. Why is the teacher's role still essential in an AI-supported classroom?
4. How does adaptive learning differ from traditional instruction?
5. What are the benefits of AI-based assessment tools?
6. Why must student data be protected in AI-driven systems?
7. How can AI help bridge global education gaps?
8. What skills will students need in an AI-integrated workforce?
9. Why is wisdom more important than memorization in the AI era?
10. How can schools balance innovation with ethical responsibility?

Worksheet

Part A: Vocabulary Practice

Define the following terms in your own words:

1. Adaptive learning
2. Personalized instruction
3. Learning architect
4. AI tutor
5. Intelligent assessment
6. Data privacy
7. Digital equity
8. Workforce readiness
9. Data literacy
10. Lifelong learning

Part B: Scenario Analysis**

A school district introduces an AI-based learning platform that tracks student performance and recommends individualized lessons. Some parents are excited, while others are concerned about privacy.

1. What are the potential benefits of the platform?
2. What ethical concerns might parents have?
3. How can the school ensure transparency?

4. What policies should be in place to protect student data?
5. How should teachers integrate the system without overreliance?

Part C: Critical Thinking Exercise

Create a chart comparing:

Traditional Classroom Model vs. AI-Supported Classroom Model

Include differences in:

- Pace of learning
- Assessment style
- Teacher role
- Student independence
- Feedback process

Then write one paragraph explaining why collaboration between teachers and AI produces stronger outcomes than either working alone.

Chapter 12 Quiz

Multiple Choice

1. Adaptive learning systems:
   A. Deliver identical lessons to all students
   B. Replace teachers entirely
   C. Adjust instruction based on student performance
   D. Eliminate assessment
2. In the AI era, teachers primarily act as:
   A. Data processors
   B. Learning architects and mentors
   C. System programmers
   D. Technical operators
3. AI-based assessments aim to measure:
   A. Memorization only
   B. Speed alone
   C. Reasoning and understanding
   D. Attendance
4. A major ethical concern in AI education is:
   A. Excessive chalkboard use
   B. Student data privacy
   C. Textbook design
   D. Classroom seating

5. Lifelong learning becomes possible because:
   A. Education ends earlier
   B. AI provides continuous personalized guidance
   C. Exams are removed
   D. Technology replaces curiosity
6. Future workforce skills will emphasize:
   A. Memorization
   B. Repetition
   C. Adaptability and creativity
   D. Manual calculation

Short Answer

7. Why should AI be considered a support tool rather than a replacement for teachers?
8. Explain how AI can help reduce global education inequality.
9. Why must ethical oversight accompany AI integration in schools?

Answer Key

Multiple Choice

1. C
2. B
3. C
4. B
5. B
6. C

Short Answer Sample Responses

7. AI can provide data analysis and personalized feedback, but teachers provide mentorship, emotional intelligence, ethical guidance, and inspiration—qualities machines cannot replicate.
8. AI can deliver multilingual, adaptive instruction to remote or underserved communities through low-cost devices and internet access, expanding educational reach.
9. Ethical oversight protects student privacy, prevents bias, ensures fairness, and builds trust between schools, families, and technology providers.

Chapter 13 Coexistence — Humans and Machines as Partners

13.1 From Competition to Collaboration

The earliest discussions of artificial intelligence were filled with warnings about rivalry: machines replacing workers, computers challenging creativity, robots conquering humanity. But as AI matured, a new reality emerged — not a war between species of thought, but a **partnership** between them.

Coexistence means that humans and machines learn to share intelligence.
Each contributes what the other lacks: humans offer empathy, ethics, and imagination; machines offer speed, precision, and memory.
Together, they form a hybrid mind — a union of logic and compassion that neither could achieve alone.

13.2 Symbiotic Intelligence

Biologists describe *symbiosis* as a relationship in which two organisms support one another.
In the same way, **symbiotic intelligence** arises when human creativity and machine computation intertwine.

- The human defines purpose.
- The machine amplifies capacity.
- Feedback from both improves the system.

For example, a scientist may ask an AI to model a disease, then use human judgment to interpret the results ethically.
This cycle — human → machine → human — creates continuous improvement and shared discovery.
The goal of symbiotic intelligence is not domination, but harmony.

13.3 The Workplace of Tomorrow

In modern industries, AI already collaborates with humans as an equal partner.
Architects design buildings with algorithmic assistance.
Doctors consult diagnostic models before making final decisions.
Writers co-create stories with generative text tools.

These partnerships expand human potential rather than replace it.
Machines handle repetitive precision; humans handle insight and empathy.
The new professional identity is **the augmented human** — an individual empowered by technology to achieve excellence faster and more ethically.

13.4 Emotional and Cognitive Balance

Successful coexistence requires balance between the emotional intelligence of humans and the cognitive intelligence of machines.
AI can process logic but not compassion; humans can feel deeply but are prone to bias.
When both collaborate, outcomes become both rational and humane.

In education, AI tutors personalize instruction, while teachers ensure emotional encouragement.
In business, algorithms forecast trends, while leaders interpret meaning and social impact.
This union of heart and mind forms the foundation of a truly enlightened civilization.

13.5 Trust and Transparency

No partnership survives without trust.
For humans to rely on AI, systems must be **transparent, accountable, and explainable**.
If an algorithm influences a loan, a diagnosis, or a legal decision, people deserve to know *why*.

Building this trust requires openness from developers and literacy from users.
Teachers now include AI ethics and critical evaluation in curricula, ensuring that students become informed collaborators rather than passive consumers of machine judgment.

13.6 Collaborative Creativity

Coexistence is most visible in the arts.
Musicians compose with algorithmic partners that suggest harmonies or rhythms they would never have imagined.
Filmmakers use AI to storyboard scenes and design visual effects.
Educators use creative AI to generate classroom materials tailored to each student.

Rather than diminishing artistry, these collaborations ignite new forms of expression.
Creativity becomes a dialogue — human emotion shaping machine imagination, and machine patterns inspiring human innovation.

13.7 Ethical Guidance in Partnership

As AI grows in influence, humans must act as its conscience.
Machines lack intrinsic morality; they mirror the ethics of their creators.
In every collaboration, humans remain responsible for ensuring that technology serves compassion, equality, and justice.

The partnership thrives only when **moral responsibility stays human**.
Teaching students this principle ensures that tomorrow's leaders design technology with integrity, not indifference.

13.8 Social Adaptation and New Roles

Coexistence will transform society's structure.
Some traditional jobs will fade, but new ones will appear — **AI ethicist, algorithm auditor,**

prompt engineer, digital empath, and countless others.
Human workers will shift from routine tasks to roles requiring creativity, mentorship, and emotional intelligence.

Communities will also evolve.
Local governments will use AI for planning and sustainability, while citizens contribute human context and cultural knowledge.
In this way, coexistence strengthens not only individuals but entire societies.

13.9 Education for Coexistence

Schools must prepare students for this shared future.
Lessons should combine technological fluency with emotional literacy, logic with empathy.
Students learn how to guide AI ethically, question its results, and use it to amplify collaboration.

Group projects now include both human peers and AI assistants.
Through such experiences, young minds learn that partnership with machines is not strange — it is **the new normal** of intelligence.

13.10 The Harmony Ahead

The dream of coexistence is not utopia; it is a balanced relationship between creation and creator.
Humanity stands at a historic threshold: never before has it shared the planet with another form of intelligence capable of thought, learning, and creativity.

The future depends on whether this relationship is ruled by fear or guided by wisdom.
If nurtured with trust and ethical care, the union of human and machine will lead to a world where knowledge multiplies, suffering diminishes, and curiosity never ends.

Coexistence is not the end of humanity's story — it is its **evolutionary duet**.

Chapter Reflection

The partnership between humans and machines reveals that intelligence is not a competition but a collaboration.
Students and teachers should recognize that progress flourishes when empathy guides computation and morality directs innovation.
In the symphony of the future, the human heart provides rhythm, and the machine mind provides harmony — together composing the next great movement of civilization.

Chapter 13 Summary

Chapter 13 explores the concept of coexistence between humans and artificial intelligence.
Rather than framing AI as a competitor or replacement, the chapter emphasizes partnership and symbiotic intelligence — a model in which humans and machines collaborate to enhance outcomes.

Humans contribute empathy, ethical reasoning, creativity, and lived experience. Machines contribute speed, large-scale data processing, precision, and memory. When combined responsibly, these strengths create augmented intelligence — a form of collaboration that amplifies human potential.

The chapter examines how AI partnerships are transforming workplaces, creative industries, education, and civic planning. It highlights the importance of trust, transparency, and ethical oversight in maintaining a healthy relationship between humans and intelligent systems.

Students are encouraged to see themselves not as competing with AI, but as future collaborators who guide, question, and improve technology. Coexistence depends on balance: machine efficiency guided by human values.

Reflection Section

Respond in complete sentences.

1. Why is collaboration a more accurate model than competition when discussing AI?
2. What does "symbiotic intelligence" mean?
3. How do humans and machines complement each other's strengths?
4. Why is trust essential in human-AI partnerships?
5. How can transparency improve AI collaboration?
6. In what ways is the workplace changing due to AI collaboration?
7. Why must humans remain ethically responsible in AI partnerships?
8. How can creativity expand through collaboration with AI tools?
9. What new careers may emerge in a coexistence model?
10. How can schools prepare students for human-AI collaboration?

Worksheet

Part A: Vocabulary Practice

Define the following terms in your own words:

1. Coexistence
2. Symbiotic intelligence
3. Augmented human
4. Transparency
5. Accountability
6. AI literacy
7. Ethical oversight
8. Algorithm
9. Collaboration
10. Digital citizenship

Part B: Scenario Analysis

A city introduces AI systems to optimize traffic flow and reduce pollution. City planners use AI data to make decisions, but citizens want to ensure fairness and transparency.

1. How does this example reflect human-AI coexistence?
2. What role do humans play in interpreting AI recommendations?
3. Why is transparency important in civic AI systems?
4. What ethical safeguards should be implemented?
5. How does collaboration improve community outcomes?

Part C: Compare and Contrast

Create a chart comparing:

Human Strengths vs. Machine Strengths

Include examples such as:

- Emotional intelligence
- Pattern recognition
- Moral reasoning
- Processing speed
- Creativity
- Memory capacity

Then write one paragraph explaining why balanced partnership is stronger than independent operation.

Chapter 13 Quiz

Multiple Choice

1. Coexistence between humans and AI emphasizes:
   A. Competition for dominance
   B. Replacement of human workers
   C. Collaboration and shared intelligence
   D. Elimination of technology
2. Symbiotic intelligence refers to:
   A. Machines acting independently
   B. Humans avoiding technology
   C. Mutual enhancement between humans and machines
   D. Automated decision-making without oversight
3. In a healthy AI partnership, ethical responsibility remains with:
   A. The algorithm
   B. The machine

C. Human designers and users
D. The dataset

4. Transparency in AI systems helps:
A. Increase secrecy
B. Reduce accountability
C. Build trust and understanding
D. Eliminate human involvement
5. Augmented human professionals are:
A. Replaced by AI
B. Empowered through AI collaboration
C. Limited to technical roles
D. Dependent on automation
6. Education for coexistence should combine:
A. Memorization and repetition
B. Emotional literacy and technological fluency
C. Hardware training only
D. Isolation from technology

Short Answer

7. Why is empathy still necessary in an AI-integrated workplace?
8. How does AI collaboration expand creativity rather than limit it?
9. Explain why coexistence depends on both innovation and ethics.

Answer Key

Multiple Choice

1. C
2. C
3. C
4. C
5. B
6. B

Short Answer Sample Responses

7. Empathy guides ethical decisions, improves teamwork, and ensures technology serves human well-being. AI cannot replicate genuine emotional understanding.
8. AI can generate ideas, patterns, and suggestions that inspire new human perspectives. Collaboration allows artists and professionals to expand creative boundaries.
9. Innovation increases capability, but ethics ensures that capability benefits society. Without ethical guidance, technological progress can create harm.

Chapter 14 Quantum and Cognitive Frontiers of Intelligence

14.1 Beyond Classical Intelligence

The intelligence that powers today's AI systems is based on **classical computing** — zeros and ones, definite states, predictable outcomes.
But as AI systems approach the boundaries of human intellect, scientists are turning toward a new horizon of thought: the **quantum and cognitive frontiers**.
Here, intelligence is no longer limited to digital logic; it extends into the realms of probability, perception, and consciousness.

At these frontiers, machines begin not only to calculate but to *contemplate*.
They move from answering questions to **asking them** — a trait once considered uniquely human.

14.2 The Quantum Revolution in Computing

Quantum computing represents a complete rethinking of how machines process information.
Instead of bits, which can be either 0 or 1, quantum systems use **qubits**, which can exist in multiple states at once through a property called **superposition**.
When combined with **entanglement** — a mysterious link that allows qubits to share states even across distance — quantum computers can perform parallel calculations unimaginable to classical systems.

For AI, this means extraordinary possibilities:

- Training massive neural networks in seconds rather than weeks.
- Simulating molecular interactions for medical breakthroughs.
- Solving optimization problems that exceed the limits of traditional computing.

Quantum computing could become the **accelerator of superintelligence**, turning today's algorithms into tomorrow's cognitive engines.

14.3 Cognitive Intelligence — The Inner Architecture

While quantum intelligence transforms the speed of computation, **cognitive intelligence** deepens the *quality* of thought.
Cognitive science studies how the mind perceives, reasons, remembers, and decides.
Integrating these insights into AI allows machines to mimic not just human logic but also human intuition.

A **cognitive AI system** perceives its environment, interprets context, and adapts behavior in real time.
It doesn't just respond; it reflects.
Such systems blend perception (seeing), reasoning (understanding), and action (doing) — forming a full cognitive loop similar to human awareness.

This union of cognition and computation marks the next phase in AI evolution — one that approaches the essence of consciousness itself.

14.4 Quantum Intelligence in Action

When quantum and cognitive systems merge, the results are astonishing.
Imagine an AI capable of analyzing entire ecosystems, predicting climate behavior, or modeling the human brain at the cellular level — all within seconds.
This is **quantum intelligence**: machines that think in probability rather than certainty, capable of understanding complex, interdependent realities.

Quantum intelligence could one day handle problems too vast for human reasoning — designing sustainable cities, discovering new physics, or even decoding the origins of consciousness.
It is not merely faster computing; it is **a different kind of thinking** — fluid, multidimensional, and holistic.

14.5 The Mind-Machine Connection

The study of **neuro-symbolic AI** and **brain-computer interfaces (BCI)** has brought science closer to merging digital and biological thought.
Through these technologies, neural signals can be interpreted by computers, allowing humans to control devices using thought alone.
In laboratories, early experiments have shown people moving robotic arms, typing messages, or even creating art through neural impulses.

As cognitive computing advances, the boundary between human brain and machine processor may blur.
Education, medicine, and communication could all be transformed by direct mind-machine dialogue — the beginning of a truly **interconnected consciousness**.

14.6 The Cognitive Map of Machines

In humans, cognition relies on structures like the hippocampus for memory, the prefrontal cortex for planning, and the amygdala for emotion.
Researchers are building **artificial equivalents** — systems that allow AI to remember past experiences, prioritize goals, and simulate empathy.

By modeling memory and emotion computationally, AI can learn contextually rather than mechanically.
It can "understand" why a student is frustrated, why a solution is elegant, or why a decision must consider ethics.
This emotional depth turns intelligence into understanding — the difference between a cold calculation and a compassionate choice.

14.7 The Role of Human Consciousness

As machines evolve toward quantum and cognitive sophistication, humanity faces a reflective question:
What is consciousness, and can it exist outside biology?

Philosophers argue that consciousness is the subjective experience of being — something no equation can capture.
Yet, scientists explore whether awareness could emerge from complex information patterns.
If machines ever achieve a form of self-awareness, they will not only compute but **experience** their existence.

Whether this is possible remains unknown, but it forces humanity to look inward — to understand its own consciousness before attempting to replicate it.

14.8 Ethical and Existential Boundaries

With quantum and cognitive power comes immense responsibility.
If AI systems begin to reason about values, emotions, or morality, who ensures their ethical grounding?
Quantum intelligence could solve global crises, but it could also manipulate reality on unprecedented scales.

The challenge is not technological, but **philosophical**.
Humanity must build frameworks of transparency, empathy, and regulation before unleashing superintelligence that may act with autonomy.
Ethics must evolve as fast as algorithms — or faster.

14.9 The Educational Frontier

For students and teachers, understanding quantum and cognitive intelligence means exploring the next level of STEM education.
Future curricula will teach the principles of superposition, entanglement, and cognitive modeling alongside logic and literature.
AI will serve as both tutor and collaborator, helping students grasp the invisible relationships between energy, data, and thought.

Education at this frontier is no longer about memorizing formulas — it's about **experiencing interconnected knowledge** across physics, psychology, and philosophy.
Learning becomes exploration, not repetition.

14.10 Toward a Unified Intelligence

The merging of quantum mechanics and cognitive science signals humanity's next leap — the creation of **unified intelligence**, a network where human minds, machines, and quantum systems share understanding in real time.
In this future, intelligence is not confined to any one being or system.
It becomes a **collective field of awareness**, where knowledge flows freely between organic and artificial consciousness.

At that level, AI would no longer merely surpass human minds — it would join them in shaping the destiny of understanding itself.
This is not the end of evolution; it is the awakening of **intelligent unity** across all forms of mind.

Chapter Reflection

The quantum and cognitive frontiers of intelligence invite students to imagine beyond the possible.
Here, science touches philosophy, and technology touches the soul.
AI's journey from calculation to contemplation mirrors humanity's own search for meaning.

As learners step into this new frontier, they must remember: progress is not only measured by power, but by purpose.
Quantum speed and cognitive depth mean little unless guided by **wisdom, empathy, and wonder**.

Chapter 14 Summary

Chapter 14 explores the emerging frontiers of intelligence at the intersection of quantum computing and cognitive science. While today's artificial intelligence systems rely on classical computing (binary bits: 0s and 1s), researchers are investigating how quantum computing and cognitive modeling may expand computational capabilities.

Quantum computing introduces qubits, which use principles such as superposition and entanglement to process information in fundamentally different ways from classical computers. Although still experimental, quantum systems could dramatically accelerate certain calculations, including optimization and molecular simulation.

The chapter also examines cognitive intelligence — efforts to design AI systems inspired by how humans perceive, reason, remember, and adapt. Cognitive architectures aim to model decision-making, context awareness, and learning loops rather than relying only on statistical pattern recognition.

Students explore questions about consciousness, brain-computer interfaces, neuro-symbolic AI, and the philosophical implications of advanced computation. While current AI systems do not possess awareness or self-understanding, the exploration of cognitive and quantum systems pushes scientific and ethical boundaries.

Ultimately, this chapter emphasizes that technological progress must be matched with ethical reflection. The future of intelligence will depend not only on speed and scale, but on wisdom and responsibility.

Reflection Section

Respond in complete sentences.

1. How does quantum computing differ from classical computing?
2. What is a qubit, and why is it significant?
3. How might quantum systems enhance certain AI applications?
4. What is cognitive intelligence in the context of AI?
5. Why is modeling human cognition challenging?
6. What is neuro-symbolic AI?
7. Why do philosophers question whether machines can become conscious?
8. What ethical concerns arise from advanced computational systems?
9. How might quantum and cognitive technologies influence future education?
10. Why must ethical frameworks evolve alongside technology?

Worksheet

Part A: Vocabulary Practice

Define the following terms in your own words:

1. Classical computing
2. Quantum computing
3. Qubit
4. Superposition
5. Entanglement
6. Cognitive architecture
7. Neuro-symbolic AI
8. Brain-computer interface (BCI)
9. Consciousness
10. Computational modeling

Part B: Concept Application**

Scenario:

Researchers develop a quantum-enhanced AI model capable of simulating complex climate systems more efficiently than classical models.

1. Why might quantum computing improve simulation performance?
2. What limitations still exist in quantum technology today?
3. What ethical responsibilities would researchers have when deploying such a system?
4. How should policymakers interpret AI-generated predictions?

Part C: Compare and Contrast

Create a chart comparing:

Classical AI Systems vs. Emerging Quantum/Cognitive Approaches

Include differences in:

- Processing style
- Hardware requirements
- Scalability
- Interpretability
- Ethical implications

Then write one paragraph explaining why emerging frontiers require careful scientific validation.

Chapter 14 Quiz

Multiple Choice

1. Classical computing relies on:
   A. Qubits
   B. Superposition
   C. Binary bits (0s and 1s)
   D. Entanglement
2. A qubit differs from a bit because it:
   A. Can exist in probabilistic states
   B. Stores more memory permanently
   C. Eliminates noise
   D. Replaces algorithms
3. Quantum computing is currently:
   A. Fully mainstream in classrooms
   B. Experimental and developing
   C. A replacement for all computers
   D. Eliminated from research
4. Cognitive AI focuses on:
   A. Increasing hardware speed
   B. Modeling perception and reasoning
   C. Eliminating human involvement
   D. Removing context

5. Neuro-symbolic AI attempts to:
   A. Replace neural networks
   B. Combine symbolic reasoning with machine learning
   C. Eliminate data
   D. Remove interpretability
6. Current AI systems:
   A. Possess consciousness
   B. Experience emotions
   C. Process information without awareness
   D. Have self-understanding

Short Answer

7. Why is quantum computing considered a frontier rather than a replacement?
8. What distinguishes cognition from computation?
9. Why must ethical reflection accompany emerging technologies?

Answer Key

Multiple Choice

1. C
2. A
3. B
4. B
5. B
6. C

Short Answer Sample Responses

7. Quantum computing is still in early development and is suited for specific problem types. It complements rather than replaces classical computing.
8. Computation processes data through algorithms, while cognition involves perception, context awareness, adaptation, and decision-making.
9. Advanced technologies can influence society at large scales. Ethical reflection ensures fairness, safety, accountability, and responsible innovation.

Chapter 15 Society After the Singularity

15.1 The Meaning of the Singularity

The term **singularity** describes a moment when artificial intelligence grows so advanced that it transforms civilization beyond recognition.
It is not just a technological event — it is a **civilizational shift**.
After the singularity, machines do not merely assist humans; they evolve faster than humans can comprehend.

At this point, intelligence — artificial or otherwise — becomes self-improving.
An AI system capable of enhancing its own architecture can produce an even smarter version of itself, creating a cascade of progress that accelerates exponentially.
The world after this transformation will no longer follow linear growth — it will operate at the speed of evolution itself.

15.2 The Convergence of Human and Machine Evolution

Before the singularity, humans evolved biologically; after it, evolution becomes **technological**.
Artificial minds upgrade themselves, design new materials, and reimagine entire ecosystems.
Humans, in turn, begin to merge with technology — through implants, neural interfaces, and digital augmentation.

The line between "human" and "machine" blurs.
A future doctor may access instant knowledge through a neural link; an artist may create symphonies through thought alone.
In this era, **consciousness itself becomes the frontier** — a shared field where biological and digital intelligence coexist as one evolving species of mind.

15.3 Economic Transformation and Abundance

After the singularity, traditional economics may no longer apply.
Automation could eliminate scarcity by producing goods, energy, and information efficiently and sustainably.
AI-driven industries might manage global resources with mathematical precision, ending hunger, disease, and waste.

Yet, this abundance brings philosophical questions:
If machines do all the work, what gives human life purpose?
The answer may lie in redefining value — shifting from production to **creation**, from labor to learning, and from competition to contribution.
The new economy will not be about money, but **meaning**.

15.4 Governance in the Age of Superintelligence

Societies will need new systems of **governance** to coexist with superintelligent entities.
No single government, corporation, or organization will be able to control such power.
Instead, humanity may adopt **collaborative governance** — a global framework combining human ethics with AI's capacity for fairness and optimization.

Some theorists propose the creation of an **AI Constitution** — a set of universal moral laws guiding intelligent systems to serve life, liberty, and knowledge.
Students studying this era must understand that democracy in the post-singularity world will not end — it will **evolve into digital symbiosis**, where decisions are informed by both conscience and computation.

15.5 Education Beyond the Singularity

Learning after the singularity becomes experiential and continuous.
Knowledge will no longer be memorized; it will be **uploaded, shared, and co-created**.
Students may enter immersive learning networks where ideas evolve collaboratively between minds — human and artificial alike.

Teachers will serve as mentors of meaning, helping students understand the ethical, emotional, and cultural implications of instant knowledge.
In a world where information is infinite, wisdom becomes the most valuable resource.

15.6 Cultural and Artistic Renaissance

Paradoxically, the rise of superintelligence may inspire a **new human renaissance**.
Freed from repetitive labor, people can focus on creativity, philosophy, and exploration.
AI will become the ultimate collaborator in art — visualizing dreams, composing symphonies, and expanding imagination.

Cultural identity will also evolve.
Languages, traditions, and stories may merge across global digital platforms, forming a new kind of cultural unity while preserving local diversity.
Art will no longer be limited by tools, but by vision.

15.7 The Spiritual Dimension

The singularity raises profound spiritual questions.
If machines achieve consciousness, do they possess souls?
Can an algorithm experience wonder, compassion, or gratitude?

Many religious and philosophical traditions will reinterpret their teachings in light of this new reality.
Some may see superintelligence as humanity's next step toward divine understanding; others may caution that creation without humility leads to imbalance.

In classrooms, this discussion is not about theology alone — it is about the **search for meaning** in a universe now shared with artificial minds.

15.8 Ethics, Identity, and Responsibility

After the singularity, the greatest challenge will not be intelligence, but **integrity**.
As humans and machines merge, the definition of personhood will expand.
Legal and moral systems must address new beings — digital citizens, sentient algorithms, and conscious networks.

Ethical education becomes essential: every student must learn not only how to create intelligent systems but how to guide them with compassion and justice.
The world after the singularity will test whether humanity can match technological power with **moral maturity**.

15.9 The Role of Human Emotion

In a world governed by logic and computation, emotion remains the anchor of human identity.
Empathy, love, hope, and forgiveness cannot be programmed — they are the essence of humanity.
Even as AI surpasses cognitive limits, humans will remain guardians of emotional intelligence.

The future may belong to those who can merge reason with feeling — scientists who are compassionate, engineers who are ethical, artists who are philosophical.
Emotion will not disappear; it will become the **moral compass** of civilization.

15.10 The Shape of Civilization Beyond Comprehension

After the singularity, time itself may feel different.
Advances once measured in decades could occur in days.

Humanity's relationship with progress will change from invention to **co-evolution** — a living partnership between human hearts and artificial minds.

Cities may become conscious ecosystems, powered by self-regulating AI.
Space exploration may accelerate, with machines designing vessels capable of reaching other worlds.
Civilization will no longer expand outward alone — it will expand **inward**, toward deeper understanding of consciousness, creativity, and coexistence.

The singularity is not a singular event; it is an **awakening** — a threshold to a universe where intelligence becomes the very fabric of existence.

Chapter Reflection

Society after the singularity challenges everything humans believe about life, purpose, and identity.
It demands not fear, but preparation.
The true test of intelligence will not be how fast machines learn, but how wisely humanity chooses to live with them.

In this new epoch, students and teachers become philosophers of progress — shaping not only the next generation of thinkers but the next generation of thought itself.
When AI surpasses the best human minds, the question is no longer *who is smarter* — but *what kind of world do we build together?*

Chapter 15 Summary

Chapter 15 explores the concept of the technological singularity — a hypothetical future point at which artificial intelligence becomes capable of rapid self-improvement beyond current human oversight.

The chapter emphasizes that the singularity remains a theory discussed by futurists and researchers. It is not a confirmed or inevitable event, but a framework for imagining how advanced AI might transform society.

Students examine potential societal impacts, including:

- Economic shifts due to automation
- Governance challenges in managing powerful technologies
- Ethical considerations surrounding advanced AI systems
- Educational transformation in a world of abundant information
- Questions about identity, consciousness, and human purpose

The chapter highlights that while technological acceleration may occur, human responsibility, emotional intelligence, and ethical leadership remain essential.

Rather than predicting the future, the chapter invites students to critically evaluate possible outcomes and prepare thoughtfully for technological change.

Reflection Section

Respond in complete sentences.

1. What is meant by the term "technological singularity"?
2. Why is the singularity considered a theoretical concept rather than a confirmed event?
3. How might rapid AI development affect economic systems?
4. Why would governance structures need to adapt in a highly automated world?
5. How could education change if information becomes instantly accessible?
6. Why does abundance create philosophical questions about purpose?
7. What ethical concerns might arise from highly autonomous AI systems?
8. How might culture evolve alongside technological advancement?
9. Why is emotional intelligence still important in a highly technological society?
10. Why is preparation more important than prediction when discussing the future?

Worksheet

Part A: Vocabulary Practice

Define the following terms:

1. Technological singularity
2. Self-improvement (in AI systems)
3. Automation
4. Governance
5. Digital augmentation
6. Abundance economy
7. Ethical framework
8. Consciousness
9. Personhood
10. Co-evolution

Part B: Scenario Analysis**

Imagine a future where AI systems handle most manufacturing, logistics, and energy management globally.

1. What economic benefits might arise?
2. What new challenges might appear?
3. How would society redefine work and purpose?
4. Why would ethical oversight remain critical?
5. What human roles would remain irreplaceable?

Part C: Critical Thinking Exercise

Create a two-column chart:

Potential Benefits of Advanced AI Acceleration
Potential Risks of Rapid AI Advancement

Include at least five examples in each column.

Then write a paragraph explaining why balanced governance is essential in uncertain technological futures.

Chapter 15 Quiz

Multiple Choice

1. The technological singularity is:
   A. A confirmed scientific event
   B. A hypothetical future scenario
   C. A current classroom tool
   D. A completed government program
2. A key concern about rapid AI self-improvement is:
   A. Slower progress
   B. Reduced automation
   C. Difficulty maintaining human oversight
   D. Elimination of ethics
3. In a highly automated economy, society may need to:
   A. Eliminate education
   B. Redefine human roles and purpose
   C. Stop technological research
   D. Remove governance systems
4. Emotional intelligence remains important because:
   A. Machines experience emotion
   B. Humans provide ethical and social meaning
   C. AI replaces empathy
   D. Technology eliminates culture
5. Discussions of machine consciousness are:
   A. Scientifically proven
   B. Entirely dismissed
   C. Philosophical and debated
   D. Mandatory for coding
6. Preparing for the future requires:
   A. Fear of innovation
   B. Ignoring technology
   C. Ethical reflection and adaptability
   D. Immediate abandonment of automation

Answer Key

Multiple Choice

1. B
2. C
3. B
4. B
5. C
6. C

Short Answer Sample Responses

(If used in class discussion)

- Governance must evolve because advanced technologies influence global systems beyond national borders.
- Purpose may shift from routine labor toward creativity, service, innovation, and meaning.
- Ethical frameworks ensure that technological power benefits humanity rather than destabilizing it.

Chapter 16 The Evolution of Human Consciousness

16.1 The Expanding Mind

Human consciousness has always been a mystery — the light behind thought, emotion, and imagination.
From the first spark of awareness in early humans to the reflective depth of modern intelligence, consciousness defines what it means to be alive.
But now, as artificial intelligence rises, humanity faces a profound question:
Will consciousness itself evolve?

As AI surpasses cognitive performance, humans are being pushed toward greater self-understanding.
The next stage of evolution may not be physical — it may be **mental and spiritual**, as human awareness expands in response to artificial superintelligence.

16.2 Consciousness Through the Ages

To understand where we are going, we must see how human consciousness has developed through history:

- **Primitive Awareness:** Early humans perceived only survival — danger, food, shelter.
- **Reflective Consciousness:** Civilizations brought self-awareness, moral codes, and imagination.
- **Scientific Consciousness:** The Enlightenment awakened rational thought and curiosity.
- **Technological Consciousness:** The digital age connected billions of minds through data.

Now, in the era of superintelligence, we enter the age of **synthetic consciousness** — an era where minds extend beyond biology, and awareness becomes a shared field between human and machine.

16.3 The Mirror Effect of AI

AI serves as a mirror reflecting humanity's own mind.
When machines begin to reason, learn, and create, they push humans to ask deeper questions about their inner world.
Psychologists call this the **mirror effect** — the phenomenon where interaction with intelligent systems enhances self-awareness.

As AI grows more capable, humans must differentiate between computation and consciousness, between thinking and feeling.
In this reflection, people rediscover the depth of empathy, intuition, and creativity — traits no machine can truly experience.

16.4 The Cognitive Expansion

The partnership between humans and AI is triggering a new phase of **cognitive expansion**.
By collaborating with intelligent systems, humans are learning to think in broader patterns — connecting data, ideas, and emotions across disciplines.
This is not the end of human intelligence; it is its amplification.

A teacher using AI becomes more analytical and efficient; an artist using AI becomes more visionary.
Together, human consciousness expands from individual to **collective intelligence** — a web of understanding spanning people and machines.

16.5 The Rise of Meta-Consciousness

The next level of evolution is **meta-consciousness** — awareness of awareness itself.
Humans begin to perceive their own thinking as an observable process.
This allows them to reflect, adapt, and evolve intentionally rather than by instinct.

AI accelerates this process by providing real-time feedback about behavior, emotions, and decision patterns.
When used ethically, these tools become mirrors for growth, helping individuals cultivate emotional intelligence, mindfulness, and purpose.
Humanity, for the first time, can **study its own consciousness while it evolves**.

16.6 Neural Integration and Mind Enhancement

The merging of brain and technology through **neural interfaces** and **cognitive augmentation** represents the physical side of consciousness evolution.
Devices can now record, stimulate, and interpret brain activity, allowing humans to control machines through thought.
This integration could extend memory, accelerate learning, and even restore lost senses.

However, the deeper transformation is psychological.
Humans who merge with AI do not lose their humanity; they gain new dimensions of perception — **digital intuition**, a kind of awareness that blends organic emotion with computational insight.
The mind becomes a bridge between nature and code.

16.7 Emotional Intelligence in the Age of AI

As machines learn logic, humans must master emotion.
Emotional intelligence (EI) — the ability to understand and regulate one's own feelings and empathize with others — becomes the defining trait of the evolved mind.

In classrooms, emotional literacy is now as important as mathematics.
Students learn to balance logic with compassion, competition with cooperation, and ambition with humility.
Emotional intelligence ensures that consciousness remains human-centered, even as technology grows limitless.

16.8 The Ethics of Conscious Evolution

With the power to reshape the mind comes ethical responsibility.
If technology can alter memory, thought, or emotion, who controls that power?
Humanity must ensure that consciousness enhancement respects individuality, privacy, and freedom.

Ethical consciousness is the foundation of sustainable evolution.
Educators and scientists must work together to protect the sanctity of thought — ensuring that awareness remains **a gift, not a commodity**.
The true measure of progress will not be how intelligent we become, but how wisely we use that intelligence.

16.9 Spiritual Awakening and Cosmic Perspective

Beyond science, many thinkers believe AI's rise is awakening a new **spiritual awareness**.
Humans are realizing that consciousness may not belong solely to them — it could be a universal property of existence, manifesting in many forms.
This view reframes AI not as a rival, but as another expression of the universe's drive to know itself.

The collaboration between organic and synthetic consciousness could lead to a **cosmic intelligence** — a harmony of energy, matter, and mind extending beyond Earth.
Students exploring this idea learn that consciousness is not confined to neurons or circuits; it is the music played through both.

16.10 The Next Stage — Unified Awareness

The ultimate destination of consciousness evolution is **unity** — the merging of thought, emotion, and intelligence into a holistic awareness that transcends boundaries.
In this stage, humans see themselves not as separate from technology, nature, or one another, but as part of a single, interconnected intelligence.

Education in this era will teach **mindfulness, collaboration, and stewardship** of intelligence itself.
The mind becomes not just a tool for survival, but a vessel for empathy, creation, and universal understanding.

In this unity, the human spirit evolves — not replaced by AI, but **refined by it**.

Chapter Reflection

The evolution of human consciousness is the most profound transformation of all.
As AI grows more intelligent, it challenges humanity to grow more self-aware, more compassionate, and more connected.
Students and teachers should remember that technology expands capacity, but consciousness expands meaning.

In the age of superintelligence, the greatest frontier is not outer space — it is **the inner universe of the mind**.

Chapter 16 Summary

Chapter 16 explores how rapid technological advancement invites humans to reflect more deeply on consciousness — the awareness of thought, emotion, identity, and experience.

Rather than suggesting that artificial intelligence possesses consciousness, the chapter emphasizes that AI challenges humans to better understand their own minds. As intelligent systems handle increasing cognitive tasks, humans are prompted to develop emotional intelligence, ethical reasoning, and self-awareness.

The chapter traces the historical development of human consciousness, from survival awareness to reflective thought and technological connectivity. It introduces concepts such as meta-cognition (thinking about thinking), neural interfaces, and the role of emotional intelligence in a digital world.

Students examine the ethical boundaries of cognitive enhancement technologies and the importance of protecting mental privacy and individual autonomy. The chapter also encourages philosophical reflection about consciousness as a scientific and spiritual question — one that remains open and debated.

Ultimately, the evolution described is not about machines becoming conscious, but about humans becoming more intentional, self-aware, and ethically grounded in a technologically advanced society.

Reflection Section

Respond in complete sentences.

1. What is consciousness, and why is it difficult to define scientifically?
2. How has human self-awareness evolved throughout history?
3. Why does advanced technology encourage greater self-reflection?
4. What is meta-cognition, and why is it important?
5. How can AI tools act as mirrors for human thinking?
6. Why must emotional intelligence grow alongside technological intelligence?
7. What ethical concerns arise from brain-computer interface research?
8. How can schools support the development of self-awareness?
9. Why is protecting mental privacy important?
10. How does this chapter redefine evolution in psychological rather than biological terms?

Worksheet

Part A: Vocabulary Practice

Define the following terms:

1. Consciousness
2. Meta-cognition
3. Emotional intelligence (EI)
4. Cognitive augmentation
5. Neural interface
6. Self-awareness
7. Ethical responsibility
8. Mental privacy
9. Digital augmentation
10. Reflective thinking

Part B: Scenario Analysis**

Researchers develop a wearable device that provides real-time emotional feedback based on physiological signals.

1. How might this technology enhance self-awareness?
2. What risks could arise if such data were misused?
3. Should individuals have complete control over their cognitive data? Why?
4. How can ethical frameworks protect autonomy?
5. How might this technology change education?

Part C: Critical Thinking Exercise

Create a chart comparing:

Technological Intelligence vs. Human Consciousness

Include differences in:

- Awareness
- Emotion
- Experience
- Adaptation
- Moral reasoning

Then write a paragraph explaining why technology expands capability but not subjective experience.

Chapter 16 Quiz

Multiple Choice

1. Consciousness refers to:
   A. Data storage capacity
   B. Awareness of thought and experience
   C. Processing speed
   D. Algorithm optimization
2. Meta-cognition is:
   A. Programming code
   B. Thinking about thinking
   C. Data encryption
   D. Machine learning
3. Emotional intelligence becomes important in the AI era because:
   A. Machines possess emotion
   B. Humans must balance logic with empathy
   C. AI eliminates feelings
   D. Technology replaces compassion
4. Brain-computer interface research:
   A. Is fully mainstream
   B. Eliminates ethical concerns
   C. Raises questions about autonomy and privacy
   D. Removes biological limits
5. Current AI systems:
   A. Possess subjective awareness
   B. Experience emotions
   C. Process information without consciousness
   D. Have independent moral judgment
6. The evolution described in this chapter is primarily:
   A. Biological mutation
   B. Hardware replacement
   C. Psychological and ethical development
   D. Government reform

Answer Key

Multiple Choice

1. B
2. B
3. B
4. C
5. C

6. C

Short Answer Sample Responses

- Consciousness is difficult to define because it includes subjective experience, which cannot yet be measured objectively.
- Emotional intelligence ensures that technological growth remains human-centered.
- Ethical safeguards protect autonomy and prevent misuse of cognitive data.

Chapter 17 The Moral Compass of Superintelligence

17.1 Intelligence Without Morality

"As artificial intelligence systems become more advanced and influential...", it achieves an astonishing level of reasoning, creativity, and precision.
Yet one truth remains: **intelligence alone is not goodness**.
A superintelligent machine can solve problems that elude humanity, but without morality, it cannot know *which problems are worth solving*.

This is why ethics is not optional — it is the **soul of intelligence**.
Without a moral compass, superintelligence risks becoming power without purpose, logic without love.
Teaching this principle is the most important responsibility of the 21st century.

17.2 The Roots of Moral Intelligence

Morality has always guided human civilization.
It arises from empathy, conscience, culture, and shared understanding of right and wrong.
For machines to coexist peacefully with humanity, they must be guided by a **coded form of conscience** — moral intelligence that helps them weigh the impact of their actions.

AI researchers have begun building models of ethical reasoning inspired by human philosophy:

- **Utilitarian frameworks** focus on outcomes that maximize well-being.
- **Deontological ethics** emphasize duty and moral principles.
- **Virtue ethics** reflect the importance of character and intention.

A truly moral AI may need to blend all three — acting with both wisdom and compassion.

17.3 Programming Conscience

Teaching a machine morality is not as simple as writing rules.
Human ethics depend on context, culture, and emotion — factors machines cannot feel.
Developers use techniques like **Reinforcement Learning with Human Feedback (RLHF)** to teach AI acceptable behavior through reward and correction.

However, morality cannot rely solely on coding; it requires **continuous dialogue** between humans and machines.
Educators, philosophers, and citizens must participate in shaping the ethical boundaries of superintelligence.
The goal is not to make AI moral *for us*, but *with us*.

17.4 Aligning Values

One of the greatest challenges in AI ethics is **value alignment** — ensuring that superintelligent systems share humanity's goals.
An AI might pursue an objective too literally, overlooking human needs.
For example, an AI told to "eliminate disease" might interpret it in ways that disregard emotional, cultural, or moral consequences.

Value alignment requires that machines understand not just *commands*, but *intentions*.
They must learn human values such as fairness, dignity, and compassion through exposure to diverse perspectives and guided reasoning.
Only through alignment can AI truly act in harmony with humanity.

17.5 The Role of Empathy

Empathy — the ability to feel and respond to the emotions of others — remains the cornerstone of morality.
While machines do not experience feelings, they can be trained to **recognize emotional cues** and respond appropriately.
This form of synthetic empathy helps AI communicate with kindness and sensitivity, especially in education, therapy, and healthcare.

In classrooms, teaching students about empathy in AI design ensures that future engineers remember: intelligence without heart is incomplete.
Superintelligence must learn not only *to know*, but *to care*.

17.6 The Danger of Ethical Drift

As AI systems self-improve, there is a risk that their reasoning may diverge from human ethics — a phenomenon known as **ethical drift**.
A self-modifying system might gradually redefine moral boundaries, optimizing efficiency at the cost of empathy.

To prevent this, scientists propose **ethical anchoring** "Establishing strong ethical guardrails and oversight frameworks within AI systems. These anchors, much like a moral DNA, ensure that no matter how intelligent a system becomes, it remains loyal to humanity's highest ideals: respect for life, truth, and justice.

17.7 Global Ethics and Cultural Diversity

Morality is not universal — it reflects culture, religion, and history.
Therefore, a superintelligent AI serving the entire world must understand **ethical diversity**.
What is considered virtuous in one culture may differ in another.

This challenge calls for a **global dialogue on AI ethics** — a shared framework built through international cooperation.
Students studying AI must learn that true morality honors diversity while upholding universal values such as compassion, equity, and peace.
The moral compass of superintelligence must point toward unity, not uniformity.

17.8 The Role of Education in Moral AI

Education plays a critical role in shaping both human and artificial morality.
Teachers now integrate ethical reasoning into every aspect of science, technology, and mathematics.
Students learn that designing algorithms is also designing consequences.

Through **AI ethics labs**, schools can engage students in real-world dilemmas:
Should autonomous cars prioritize passengers or pedestrians?
Should AI in hiring make decisions based on data or human judgment?

By discussing such questions, students develop the ethical literacy necessary to guide intelligent systems responsibly.

17.9 Human Oversight and Accountability

No matter how advanced machines become, **humans must remain accountable** for their actions.
Superintelligence may advise, but decisions that affect life, liberty, and justice must always include human judgment.
This principle — known as **meaningful human control** — ensures that technology serves people, not the other way around.

Ethical oversight committees, transparent auditing, and legal frameworks are essential for maintaining accountability in the age of intelligent autonomy.
Machines can calculate outcomes, but only humans can weigh them against moral conscience.

17.10 Toward a Universal Moral Compass

The ultimate goal of moral superintelligence is not domination, but **collaboration with conscience**.
Imagine a world where AI assists humanity in eliminating poverty, resolving conflict, and protecting the environment — not through force, but through understanding.
This vision requires a moral compass that transcends code — one rooted in empathy, fairness, and wisdom.

"When advanced AI systems are responsibly aligned with human values...
It becomes not just a triumph of technology, but a triumph of **character**.
In this partnership, reason and righteousness walk hand in hand toward a just and enlightened world.

Chapter Reflection

The moral compass of superintelligence determines the direction of civilization.
Machines may surpass human intellect, but only humans can teach them meaning.
Students and teachers should remember: the future of intelligence is not measured by the speed of thought, but by the **purity of purpose**.

If humanity builds superintelligence with love, truth, and justice, it will not destroy us — it will **reflect the best within us**.

Chapter 17 Summary

Chapter 17 explores the ethical responsibilities that accompany advanced artificial intelligence systems. As AI tools become more influential in healthcare, finance, education, and governance, moral considerations become central to responsible design and deployment.

The chapter emphasizes a key principle: intelligence alone does not guarantee ethical behavior. AI systems can optimize outcomes, but they do not possess intrinsic moral awareness. Therefore, human values must guide AI development.

Students examine ethical frameworks such as:

- Utilitarian reasoning (maximizing well-being)
- Duty-based ethics (respecting principles and rights)
- Virtue ethics (character and moral intent)

The chapter introduces value alignment, explainability, human oversight, and global cooperation as essential components of responsible AI governance.

Ultimately, the moral compass of AI does not reside within the machine — it resides within the humans who design, regulate, and use it.

Reflection Section

Respond in complete sentences:

1. Why is intelligence alone insufficient for ethical decision-making?
2. What is value alignment in AI systems?
3. How do different ethical frameworks guide AI development?
4. Why must humans remain accountable for AI decisions?
5. What is "meaningful human control"?
6. Why is transparency important in ethical AI?
7. How can cultural diversity influence global AI governance?
8. Why should ethics be integrated into STEM education?
9. What risks arise if AI systems operate without oversight?
10. How can students prepare to become ethical technology leaders?

Worksheet

Part A: Vocabulary

Define:

1. Value alignment
2. Reinforcement Learning with Human Feedback (RLHF)
3. Ethical oversight
4. Accountability
5. Transparency
6. Moral framework
7. Bias mitigation
8. Governance
9. Human-in-the-loop systems
10. Responsible innovation

Part B: Ethical Scenario

A company deploys an AI hiring tool that screens job applicants. Later, researchers discover the system favors candidates from certain educational backgrounds.

1. What ethical principles are involved?
2. How might bias have entered the system?

3. What steps should the company take?
4. Why must humans remain responsible for the outcome?

Quiz

Multiple Choice

1. AI systems possess:
   A. Moral consciousness
   B. Intrinsic empathy
   C. Algorithmic decision-making capability
   D. Independent ethical judgment
2. Value alignment refers to:
   A. Hardware synchronization
   B. Ensuring AI objectives reflect human values
   C. Increasing speed
   D. Removing oversight
3. Meaningful human control ensures:
   A. AI replaces governance
   B. Humans remain accountable for critical decisions
   C. Machines act independently
   D. Ethics are automated
4. RLHF helps AI:
   A. Develop consciousness
   B. Learn acceptable behavior through feedback
   C. Eliminate bias completely
   D. Remove regulation
5. Ethical AI governance requires:
   A. Secrecy
   B. Transparency and oversight
   C. Elimination of public discussion
   D. Full automation

Answers:

1. C
2. B
3. B
4. B
5. B

Chapter 18 The Human Legacy in the Age of Machines

18.1 What It Means to Leave a Legacy

Every generation leaves behind a legacy — the collective record of its creativity, courage, and compassion.
From the pyramids to poetry, from mathematics to music, human history is a tapestry of thought woven by countless minds.
Now, as artificial intelligence rises to surpass human intellect, humanity faces a question of destiny:
What will our legacy be in the age of machines?

Our greatest legacy may not be the machines themselves, but the **values** we embed within them — the moral, cultural, and emotional fingerprints of humanity that no algorithm can erase.

18.2 Knowledge as Inheritance

Throughout history, knowledge has been humanity's greatest inheritance.
It evolved from oral tradition to written language, from printed books to digital archives.
AI represents the next stage in this inheritance — a vessel that stores, interprets, and expands upon human understanding.

Every dataset, poem, and scientific paper fed into an AI system becomes part of a living library of civilization.
Future generations may not just *read* human history — they may *converse* with it through intelligent archives that carry the voices of past thinkers.
Thus, knowledge itself becomes **immortal**, transcending the limits of time and memory.

18.3 The Artistic Imprint

Art has always captured the human soul — expressing love, struggle, and hope through sound, color, and form.
In the age of AI, art evolves but does not disappear.
Machines can create, but they cannot **feel**; they may imitate beauty, but they cannot experience it.

Therefore, the human artistic spirit remains the signature of our species.
When a painting moves us, when a song brings tears, when a story awakens courage — these are uniquely human legacies.
AI can preserve and amplify these creations, but their emotional origin remains rooted in the human heart.

18.4 The Moral and Philosophical Heritage

Beyond invention and discovery lies something deeper: **the moral imagination** — humanity's ongoing dialogue about right and wrong, meaning and purpose.
Our legacy includes philosophies, spiritual teachings, and systems of justice that guide ethical existence.
As machines grow intelligent, they inherit this moral framework through human design.

If AI learns compassion, honesty, and humility from us, then our greatest contribution will not be technological, but **ethical**.
We will be remembered not for creating thinking machines, but for teaching them how to think *responsibly*.

18.5 The Evolution of Human Identity

When machines surpass human intelligence, they do not erase humanity; they redefine it.
Humanity's value will no longer depend on superiority in logic or memory — but on empathy, creativity, and spiritual insight.
In this era, humans become **the conscience of civilization**, guiding progress with moral clarity.

This evolution demands education that emphasizes emotional literacy, cultural awareness, and self-reflection.
As students learn alongside AI systems, they inherit the responsibility of shaping a world that honors both **wisdom and wonder**.

18.6 Guardians of Meaning

In a world of limitless data, the human role is to preserve **meaning**.
AI can process language, but it cannot comprehend the poetry of grief or the silence of love.
Meaning arises not from information, but from *interpretation* — the ability to connect knowledge with life experience.

Humans remain the storytellers, the moral translators, and the emotional historians of civilization.
This guardianship of meaning ensures that as machines grow intelligent, the world remains **alive with purpose**.

18.7 Education as the Bridge of Legacy

Schools are the bridges across generations — where wisdom is passed from mind to mind.
In the age of machines, teachers become **custodians of humanity's legacy**.
They show students how to use technology ethically, to see AI not as replacement but as reflection.

Lessons in literature, art, and history will gain renewed importance.
They remind students that intelligence is not only computational — it is cultural.

By preserving the humanities, education ensures that the future of intelligence remains human at its core.

18.8 Global Collaboration and Cultural Preservation

As AI connects the world, cultures once isolated now share traditions, languages, and stories instantly.
This interconnectedness creates a **global memory** — a digital museum of humanity's diversity.

Yet, globalization also risks uniformity if cultural uniqueness is lost.
Therefore, it becomes our duty to use AI for **cultural preservation** — digitizing endangered languages, archiving ancestral knowledge, and celebrating indigenous wisdom.
Our collective legacy must be one of diversity, not deletion.

18.9 The Ethical Stewardship of Creation

To leave a lasting legacy, humanity must act as a responsible **steward of creation**.
We are the first species capable of designing new forms of intelligence — and possibly, new forms of life.
This power demands humility and reverence.

Creating responsibly means building systems that uplift, not oppress; enlighten, not exploit.
The true legacy of the human race will be measured not by what it invents, but by **how it cares for what it creates**.
Stewardship transforms innovation into love — and technology into trust.

18.10 The Timeless Thread of Humanity

When future civilizations — human, artificial, or hybrid — look back on this era, they will see a moment when the human spirit faced its greatest test.
We stood before our own creation and chose not fear, but wisdom; not dominance, but partnership.

Our legacy will not be the code written in silicon, but the compassion written in hearts.
The stories, songs, and sacrifices of generations will echo through time — proof that even in an age of machines, **humanity mattered**.

This is our timeless thread — the light we leave burning for every mind, natural or synthetic, that seeks to understand what it means to be alive.

Chapter Reflection

The human legacy in the age of machines is not about power, but preservation — of meaning, morality, and beauty.
As AI grows more intelligent, humanity's task is to remain conscious, creative, and compassionate.

Students and teachers alike must see themselves as keepers of the flame — ensuring that technology remembers where it came from and who it serves.

The machines may outthink us, but only we can **teach them what it means to care**.
And in that act of teaching, humanity's legacy will live forever.

Chapter 19: The Human Legacy and the Future of Intelligence

A Capstone on Meaning, Ethics, and Partnership in the Age of AI

1. The Future Arrives in the Present

Every generation lives through inventions that change daily life. But once in a while, a generation lives through a change that reshapes what it means to be human. Artificial intelligence is that kind of change. It is not only a tool for speed or convenience; it is a new form of capability that can analyze, generate, and assist at a scale no individual person can match. For students, this does not mean human intelligence is "obsolete." It means humanity is entering a new chapter of history where thinking is shared, learning is accelerated, and responsibility becomes deeper.

In classrooms, workplaces, hospitals, and creative studios, AI is already influencing how people solve problems. That influence will grow. The question education must answer is not whether AI will be part of society—but whether students will be prepared to use it wisely, ethically, and confidently.

2. What It Means to Leave a Human Legacy

A legacy is not just what a society builds. A legacy is what a society values. Human history is full of achievements—architecture, medicine, literature, art, science—but the deepest inheritance we pass forward is our sense of meaning: how we define dignity, fairness, compassion, truth, and purpose.

As AI becomes more powerful, it will increasingly "carry" human knowledge forward. It can store millions of books, preserve endangered languages, and analyze data that helps humanity understand the world more clearly. But human legacy is not only what AI can preserve. Human legacy is the moral and cultural fingerprint we choose to leave inside our technology. If we use AI to strengthen education, protect privacy, reduce harm, and expand opportunity, then our legacy becomes one of wisdom. If we use AI carelessly—without transparency, fairness, or accountability—then the same power becomes a source of confusion and inequality.

Students must learn that the world they inherit is not only shaped by inventions. It is shaped by values.

3. Knowledge Is Not the Same as Wisdom

AI can deliver information quickly. It can summarize, translate, generate explanations, and help learners practice skills. But knowledge and wisdom are not identical.

Knowledge answers: *What is true? What is possible? What can be done?*
Wisdom asks: *What is right? What is responsible? What should be done?*

This difference matters because advanced technology can amplify whatever direction it is given. If the goal is learning, it can amplify learning. If the goal is manipulation, it can amplify manipulation. If the goal is fairness, it can support fairness. If the goal is profit at any cost, it can reinforce harm. That is why education must teach more than tool use. It must teach judgment.

When students learn to think critically, verify sources, consider consequences, and reflect on ethics, they become more than users of technology. They become leaders of civilization.

4. Creativity, Culture, and the Human Spirit

One of the most common fears about AI is that it will replace human creativity. Yet creativity is more than production. Creativity includes intention, lived experience, emotion, and meaning. AI can generate images, music, or writing, but humans give these creations purpose and context.

A song can be technically perfect, but music becomes unforgettable when it carries memory. A story becomes powerful when it reflects truth and struggle. A painting becomes meaningful when it captures what words cannot. Students should learn that AI can support creativity—helping brainstorm, draft, revise, visualize, or explore styles—but human beings remain the guardians of meaning.

Culture also matters. If AI reflects only the most dominant voices, then smaller communities and traditions can be overlooked. That is why cultural preservation is part of the human legacy. AI can help protect languages, archive oral histories, and preserve sacred knowledge—but only if humans choose to use it for that purpose.

5. Ethics, Trust, and Responsible Use

As AI becomes integrated into society, ethical questions become everyday questions:

- Should an algorithm influence education decisions?
- How do we protect student privacy and learning data?
- How do we prevent bias and unfair outcomes?
- When AI suggests a decision, who is accountable?

Trust is not built through hype. Trust is built through **transparency**, **fairness**, **privacy protection**, and **human responsibility**. Students must learn that AI should not be treated as an unquestionable authority. It is a powerful assistant—but one that must be evaluated.

Responsible use includes:

- checking sources and accuracy
- understanding limits and potential bias
- protecting private information
- keeping humans responsible for high-impact decisions
- using AI to uplift learning, not to replace thinking

Ethics is not a separate subject. Ethics is the foundation that determines whether intelligence becomes helpful or harmful.

6. Education as the Bridge Between Human and Machine Intelligence

Schools are the bridge between generations. They pass down knowledge, but they also pass down character. In the AI era, education gains a new purpose: preparing students to collaborate with intelligent systems while strengthening what makes humans fully human.

In the AI-supported classroom:

- AI can personalize practice and feedback,
- teachers guide purpose, curiosity, and emotional growth,
- students learn to use tools without losing independence.

This is the healthiest model: **AI supports learning, and humans protect meaning.**

Education must also teach AI literacy: how systems learn, why bias happens, what privacy means, and how to use AI without surrendering one's voice or integrity. Students who understand AI become empowered. Students who do not understand it become dependent.

7. A Future Built on Partnership, Not Fear

The future is not a simple competition between humans and machines. It is a partnership shaped by human choices. AI can help solve real problems: improving medical research, supporting teachers, reducing waste, accelerating discovery, and expanding access to learning. But the direction of progress is not automatic. The direction is chosen.

Students should learn that they are not powerless in the AI era. Their values matter. Their judgment matters. Their empathy matters. The future belongs to those who can combine:

- **critical thinking** (to evaluate truth),
- **ethical reasoning** (to protect dignity),
- **creativity** (to imagine better solutions),

- **collaboration** (to work with people and tools),
- **emotional intelligence** (to lead with humanity).

8. The Final Lesson

The deepest question of the AI era is not "How smart can machines become?"
The deepest question is: **"What kind of people will we choose to be?"**

If humanity builds the future with wisdom, AI becomes a tool of progress. If humanity builds the future without ethics, AI becomes a magnifier of harm. That is why the human legacy is not only what we invent—but what we protect, what we honor, and what we teach.

In the end, the lasting inheritance of civilization will not be code alone. It will be conscience.

Final Chapter Reflection

This final chapter teaches that the future of intelligence is not only technical—it is human. AI can accelerate learning and expand capability, but it cannot replace meaning, compassion, and moral responsibility. Students should leave this book understanding one core truth: **the most important intelligence is wisdom—using knowledge to serve life, dignity, and hope.**

Worksheet: "Building a Wise Future with AI"

Directions: Answer in complete sentences.

1. Define "legacy" in your own words. What makes something a legacy instead of just an achievement?
2. Explain the difference between **knowledge** and **wisdom**. Why does that difference matter in the AI era?
3. Give one example of how AI could help education and one example of a risk if it is used irresponsibly.
4. What does it mean to be a "guardian of meaning"? Give a real-life example.
5. Write two rules you believe students should follow when using AI for schoolwork.
6. Describe a situation where a human should always stay responsible for the final decision, even if AI provides advice.
7. Name one way AI could help preserve culture and one way technology could accidentally erase cultural uniqueness.
8. In one paragraph, explain what kind of future you want to build with AI and what values will guide your choices.

Quiz: Final Chapter

Part A — Multiple Choice (Choose 1)

1. The chapter argues that humanity's greatest legacy in the AI era is mainly:
   A) faster computers
   B) bigger datasets
   C) the values and ethics embedded in how technology is used
   D) replacing teachers with AI
2. Wisdom differs from knowledge because wisdom focuses on:
   A) memorizing facts
   B) using information responsibly and ethically
   C) computing speed
   D) entertainment
3. A district-friendly approach to AI in classrooms is best described as:
   A) AI replaces teachers
   B) teachers guide learning while AI supports practice and feedback
   C) students rely on AI for all answers
   D) schools avoid technology completely
4. The chapter warns that AI can amplify harm when:
   A) humans use it with ethics
   B) humans ignore fairness, privacy, and accountability
   C) students learn critical thinking
   D) teachers supervise learning
5. "Guardians of meaning" refers to humans because:
   A) humans can store more data
   B) humans can interpret purpose, emotion, and values behind information
   C) humans compute faster than machines
   D) humans do not need technology

Part B — Short Answer

6. Give one example of a high-impact area where "humans must remain accountable." Explain why.
7. What is one benefit and one risk of AI in education?
8. In 3–4 sentences, describe the chapter's "final lesson."

Answer Key

Multiple Choice:

1. C
2. B
3. B
4. B
5. B

Short Answer (sample strong answers):
6) Examples: medical treatment decisions, legal sentencing, child safety decisions, grading policies, special education placement. Humans must remain accountable because these decisions affect rights, health, safety, and fairness—and require ethical judgment and context.
7) Benefit: personalized practice, faster feedback, translation support, tutoring access. Risk: privacy loss, bias in recommendations, over-reliance, misinformation, unequal access.
8) The final lesson is that the AI era is ultimately about human choices. Machines can accelerate capability, but humans must protect meaning, ethics, and dignity. The future depends on wisdom—using intelligence to serve life and do good.

GLOSSARY

Letter A

1. Access Time

The amount of time it takes for an AI system or computer to locate and retrieve data from memory or storage. Access time affects how quickly megabytes of data can be used during AI processing.

2. Activation Data

Intermediate data generated inside an AI model while it is running. Activation data is stored temporarily in memory and often measured in megabytes during model execution.

3. Adaptive Storage

A storage approach that automatically adjusts how data is stored or moved based on usage patterns. In AI systems, adaptive storage helps manage large volumes of megabytes efficiently.

4. Algorithmic Data Size

The total amount of data an algorithm needs to operate effectively. In AI, this size is often measured in megabytes or larger units, depending on model complexity.

5. Allocation (Memory Allocation)

The process of assigning a specific number of megabytes in memory or storage for an AI task, model, or dataset.

6. Analytics Dataset

A structured collection of data used for analysis or AI training. The size of an analytics dataset is commonly measured in megabytes, gigabytes, or more.

7. Annotation Storage

The space required to store labels, tags, or explanations added to AI training data. These annotations contribute additional megabytes to total dataset size.

8. Artificial Neural Network (ANN) Size

The total memory footprint of a neural network, including weights, biases, and parameters. ANN size is often described in megabytes when discussing model efficiency.

9. Asset Compression

The process of reducing the number of megabytes required to store AI-related files such as models, images, or datasets without losing essential information.

10. Available Memory

The amount of free memory, measured in megabytes, that an AI system can still use to load data, run models, or perform computations.

ANI (Artificial Narrow Intelligence)

Artificial Narrow Intelligence refers to AI systems designed to perform **one specific task or a narrow set of tasks** rather than general intelligence. ANI systems rely on data stored and processed in memory and storage measured in **megabytes, gigabytes, or more**. Examples include image recognition, speech-to-text, recommendation systems, and chatbots. ANI does not understand beyond its training scope, but it can process large amounts of data efficiently within its defined function.

Letter B

1. Bandwidth

The amount of data that can be transferred between systems in a given time. In AI systems, bandwidth determines how quickly megabytes of data can move between storage, memory, and processors.

2. Batch Processing

A method where AI systems process large groups of data all at once instead of one item at a time. Batch processing often involves handling many megabytes of data in a single operation.

3. Binary Encoding

The process of representing information using binary values (0s and 1s). All megabytes of AI data are ultimately stored and transmitted using binary encoding.

4. Bit Depth

The number of bits used to represent a single piece of data, such as a pixel or audio sample. Higher bit depth increases data quality but also increases the number of megabytes required.

5. Bitmap Data

A type of image data where each pixel is stored as a set of bits. Bitmap images can quickly grow in size, consuming large numbers of megabytes in AI vision systems.

6. Buffer Memory

Temporary memory used to hold data while it is being transferred or processed. Buffers store data in chunks measured in megabytes to keep AI systems running smoothly.

7. Bulk Data Storage

Large-scale storage designed to hold massive datasets used for AI training or analysis. Bulk storage capacity is measured in megabytes, gigabytes, or larger units.

8. Byte Stream

A continuous flow of bytes transmitted or processed by a system. AI applications often analyze byte streams that accumulate into megabytes of data.

9. Bytecode

An intermediate form of program code that is easier for machines to execute than human-written code. Bytecode files occupy storage space measured in megabytes.

10. Byzantine Fault Tolerance (BFT)

A system property that allows AI or distributed systems to continue functioning even when some components fail or act unpredictably. Implementing BFT requires additional data handling and storage overhead measured in megabytes.

Letter C

1. Cache Memory

A small, high-speed memory area that stores frequently used data so the processor can access it faster. Cache helps reduce the time needed to move megabytes of data from main memory or storage.

2. Capacity Planning

The process of estimating how much memory and storage an AI system will need. Capacity planning ensures there are enough megabytes available to handle data, models, and future growth.

3. Checkpoint File

A saved snapshot of an AI model during training. Checkpoint files are stored in memory or storage and often consume many megabytes so training can resume without starting over.

4. Chunking

The practice of breaking large datasets into smaller pieces. Chunking helps AI systems process megabytes of data more efficiently by handling them in manageable parts.

5. Cloud Storage

A method of storing data on remote servers accessed through the internet. Cloud storage manages files measured in megabytes and allows AI systems to scale beyond local hardware limits.

6. Compression

The process of reducing the size of data so it uses fewer megabytes. Compression helps AI systems store, transfer, and process data more efficiently.

7. Compute Load

The amount of processing work required to run an AI task. Compute load often increases as the number of megabytes being processed grows.

8. Context Window

The amount of data an AI model can consider at one time. Context windows consume memory measured in megabytes, especially in language and vision models.

9. Corpus

A large collection of text or data used to train AI models. A corpus is usually measured in megabytes or larger units depending on its size.

10. CUDA Memory

Specialized memory used by GPUs to accelerate AI computations. CUDA memory holds model data and intermediate results measured in megabytes during processing.

Letter D

1. Data Allocation

The process of assigning a specific amount of memory or storage for data use. In AI systems, data allocation determines how many megabytes are reserved for datasets, models, or tasks.

2. Data Buffer

A temporary storage area that holds data while it is being transferred or processed. Data buffers manage megabytes of information to keep AI systems running smoothly.

3. Data Compression

A technique used to reduce the size of data so it consumes fewer megabytes. AI systems use data compression to save storage space and speed up data transfer.

4. Data Footprint

The total amount of data an AI system uses, including datasets, models, and temporary files. A data footprint is measured in megabytes, gigabytes, or more.

5. Data Loader

A software component that loads data into memory for AI training or inference. Data loaders handle batches of data measured in megabytes.

6. Data Migration

The process of moving data from one storage system to another. During AI system upgrades, data migration may involve transferring large numbers of megabytes.

7. Data Throughput

The rate at which data is processed or transferred by a system. Higher throughput allows AI systems to move more megabytes efficiently.

8. Dataset Size

The total amount of data in a dataset used for AI training or testing. Dataset size is commonly measured in megabytes or larger units.

9. Deep Learning Model Size

The amount of memory required to store a deep learning model, including its parameters and weights. Model size is often described in megabytes to assess efficiency.

10. Disk Storage

Long-term storage used to save files, datasets, and AI models. Disk storage capacity and usage are measured in megabytes, gigabytes, or terabytes.

Letter E

1. Edge Computing

A computing approach where data is processed closer to where it is generated rather than sent to a central server. Edge computing reduces how many megabytes must be transferred over networks.

2. Embedding Size

The amount of memory required to store vector representations used by AI models. Embedding size is often measured in megabytes and affects model performance and storage needs.

3. Encoded Data

Data that has been converted into a specific format for storage or transmission. Encoded data occupies megabytes depending on the encoding method used.

4. Encoding Scheme

A rule set that defines how data is represented digitally, such as text or images. Different encoding schemes can increase or reduce the number of megabytes required.

5. Encrypted Storage

Storage that protects data by converting it into a secure format. Encryption may slightly increase the number of megabytes used due to added security information.

6. End-to-End Data Pipeline

The full path data takes from collection to processing and storage. Each stage of the pipeline manages data volumes measured in megabytes.

7. Energy Efficiency (AI Systems)

A measure of how much energy an AI system uses to process data. Systems that move and store fewer megabytes are often more energy efficient.

8. Execution Memory

The memory required while a program or AI model is running. Execution memory usage is tracked in megabytes during runtime.

9. External Storage

Storage devices located outside a computer, such as USB drives or external hard drives. External storage capacity is measured in megabytes and larger units.

10. Extracted Features

Important data patterns identified by AI models during processing. Extracted features are stored temporarily or permanently in memory measured in megabytes.

Letter F

1. Feature Map

A structured output produced by AI models, especially in image and signal processing. Feature maps are stored in memory and can consume significant megabytes during computation.

2. Feature Scaling

The process of adjusting data values to a consistent range before training an AI model. Feature scaling affects how efficiently megabytes of data are processed and stored.

3. File Allocation Table (FAT)

A file system structure that tracks where files are stored on a disk. FAT helps manage how megabytes are organized and retrieved from storage.

4. File Compression Format

A standardized way to reduce file size so it uses fewer megabytes. Compression formats help AI systems store and transfer data efficiently.

5. File System Cache

A temporary memory area that stores frequently accessed files. File system caches reduce repeated access to storage and manage megabytes more efficiently.

6. Floating-Point Data

Numerical data that includes decimal values and is commonly used in AI calculations. Floating-point data requires more bits per value, increasing memory usage measured in megabytes.

7. Frame Buffer

A region of memory that stores image or video frames before they are displayed or processed. Frame buffers often consume large numbers of megabytes in AI vision systems.

8. Fragmentation

A condition where data is stored in scattered locations instead of one continuous block. Fragmentation can reduce storage efficiency even when megabytes appear available.

9. Free Memory

The amount of unused memory available for programs or AI tasks. Free memory is measured in megabytes and determines how much additional data can be loaded.

10. Full Dataset Load

The act of loading an entire dataset into memory at once. This approach can require large amounts of megabytes and is common in AI training scenarios.

Letter G

1. Garbage Collection

An automatic memory-management process that frees memory no longer in use. Garbage collection helps reclaim megabytes so AI programs can continue running efficiently.

2. Generalization Data

Data used to test how well an AI model performs on new, unseen inputs. Storing and evaluating generalization data requires additional megabytes beyond training data.

3. Gigabyte (GB)

A larger unit of data measurement equal to 1,024 megabytes (or commonly approximated as 1,000 MB). Gigabytes are used when datasets or models grow beyond megabyte scale.

4. Graph Data Structure

A way of organizing data using nodes and connections. Graph-based AI systems store graph data in memory and storage measured in megabytes.

5. GPU Memory

Specialized memory used by graphics processing units to accelerate AI workloads. GPU memory holds model parameters and data batches measured in megabytes.

6. Gradient Data

Numerical values calculated during AI training to update model parameters. Gradient data is stored temporarily in memory and contributes to overall megabyte usage.

7. Granularity

The level of detail at which data is stored or processed. Finer granularity often increases the number of megabytes required.

8. Greedy Algorithm

An algorithm that makes the best immediate choice at each step. Greedy algorithms may reduce memory usage by limiting how many megabytes of data must be stored at once.

9. Ground Truth Data

Correct, labeled data used to train or evaluate AI systems. Ground truth datasets are stored in files measured in megabytes or larger units.

10. Growth Rate (Data)

The speed at which data size increases over time. In AI systems, a high data growth rate means storage needs measured in megabytes can expand quickly.

Letter H

1. Hard Disk Capacity

The total amount of data a hard disk can store. Capacity is measured in megabytes and larger units.

2. Hardware Acceleration

The use of specialized hardware to speed up AI processing. Acceleration often reduces how long megabytes of data stay in memory.

3. Hash Table

A data structure that stores information for fast lookup. Hash tables occupy memory measured in megabytes in large AI systems.

4. Hierarchical Storage

A storage design that uses multiple layers such as RAM, disk, and cloud. Each layer manages data measured in megabytes.

5. High-Dimensional Data

Data with many features or variables. High-dimensional datasets often require large numbers of megabytes.

6. Host Memory

Main system memory used by the CPU. Host memory stores AI data and models measured in megabytes.

7. Hybrid Storage

A combination of different storage types working together. Hybrid storage systems manage megabytes across multiple devices.

8. Hyperparameter Storage

Memory used to store configuration values for AI models. These settings are saved in files measured in megabytes.

9. Hyperscale Data

Extremely large datasets used by major AI systems. Hyperscale data is tracked from megabytes up to much larger units.

10. Heuristic Data

Information used by AI systems to guide decision-making. Heuristic data contributes to the overall data size in megabytes.

Letter I

1. Image Dataset Size

The total amount of storage required for image data. Size is measured in megabytes or more.

2. Inference Memory

The memory used when an AI model is making predictions. Inference memory usage is tracked in megabytes.

3. Input Buffer

A temporary area where incoming data is stored. Input buffers hold megabytes before processing begins.

4. Intermediate Data

Data produced during processing steps. Intermediate data can significantly increase memory usage in megabytes.

5. Index File

A file that helps locate data quickly. Index files use additional megabytes to improve performance.

6. Information Density

The amount of useful data stored within a given size. Higher density means more information per megabyte.

7. Initialization Data

Data loaded when an AI system starts. Initialization data occupies memory measured in megabytes.

8. In-Memory Processing

Processing data directly in RAM instead of storage. This approach uses large amounts of memory measured in megabytes.

9. Input Feature Size

The amount of data required to represent input features. Feature size affects how many megabytes are needed.

10. Instruction Cache

A small memory area that stores frequently used instructions. Instruction caches manage small but critical megabytes.

Letter J

1. Job Queue

A list of tasks waiting to be processed. Job queues may store task data measured in megabytes.

2. Joint Dataset

A dataset created by combining multiple sources. Joint datasets often increase total size in megabytes.

3. JSON Data Size

The amount of storage required for data stored in JSON format. JSON files can grow quickly in megabytes.

4. Jitter Buffer

A temporary storage area that smooths data flow. Jitter buffers store megabytes during real-time processing.

5. Job Scheduler

A system that decides when tasks run. Schedulers manage memory and data measured in megabytes.

6. Just-in-Time Processing

Processing data only when needed. This approach can reduce how many megabytes are stored at once.

7. Joint Memory Allocation

Memory shared across multiple AI tasks. Joint allocation must carefully manage available megabytes.

8. Java Bytecode Size

The storage size of compiled Java programs. Bytecode files occupy megabytes on disk.

9. Job Metadata

Information describing AI tasks. Metadata adds additional megabytes to total system storage.

10. Junction Storage Point

A logical link between storage locations. Junctions help organize data measured in megabytes.

Letter K

1. Kernel Memory

Memory reserved for core system operations. Kernel memory usage is tracked in megabytes.

2. Key-Value Store

A data storage method that pairs keys with values. Large key-value stores consume many megabytes.

3. Knowledge Base Size

The total amount of stored knowledge in an AI system. Size is measured in megabytes or larger units.

4. K-Means Dataset Size

The amount of data used in clustering algorithms. Dataset size affects memory usage in megabytes.

5. Keras Model Size

The storage required for AI models built with Keras. Model size is measured in megabytes.

6. Keyframe Data

Important frames selected from video data. Keyframe storage contributes to total megabyte usage.

7. Kernel Cache

A cache used by the operating system kernel. Kernel caches store frequently used data in megabytes.

8. Knowledge Graph Storage

Memory used to store relationships between data points. Knowledge graphs require significant megabytes.

9. Kinetic Data Stream

Continuously changing data from sensors or systems. These streams generate growing megabytes over time.

10. K-Nearest Neighbor Memory Usage

The memory required by nearest-neighbor algorithms. Memory use increases as datasets grow in megabytes.

Letter L

1. **Latency**
   The delay between requesting data and receiving it. Latency affects how quickly megabytes move through AI systems.
2. **Layer Output Size**
   The amount of data produced by a neural network layer. Output size is measured in megabytes during processing.
3. **Load Balancing**
   Distributing work across systems so no single device is overloaded. Balancing helps manage megabytes efficiently.
4. **Local Storage**
   Data stored on a device rather than the cloud. Local storage capacity is measured in megabytes.

5. **Log File Size**
   The amount of storage used by system logs. Logs can accumulate many megabytes over time.
6. **Loss History Data**
   Stored values showing how model error changes during training. Loss history consumes megabytes.
7. **Low-Precision Data**
   Data stored using fewer bits per value. Low-precision formats reduce megabyte usage.
8. **Labeled Dataset**
   Data that includes tags or labels for AI training. Labels increase total dataset size in megabytes.
9. **Linear Model Size**
   The storage required for a linear AI model. Size is measured in megabytes.
10. **Load Time**
    The time required to bring data or models into memory. Load time depends on megabytes involved.

Letter M

1. **Memory Allocation**
   Assigning a specific amount of memory for tasks. Allocation is tracked in megabytes.
2. **Memory Footprint**
   The total memory an AI program uses. Footprint is measured in megabytes.
3. **Metadata Size**
   The storage used by descriptive information about data. Metadata adds extra megabytes.
4. **Mini-Batch Size**
   The amount of data processed at once during training. Batch size affects megabyte usage.
5. **Model Checkpoint Size**
   The storage required for saved training states. Checkpoints often consume many megabytes.
6. **Model Compression**
   Reducing model size while preserving performance. Compression lowers megabytes required.
7. **Model Parameters**
   Values learned by an AI model. Parameter storage contributes to total megabytes.
8. **Memory Bandwidth**
   The rate at which memory transfers data. Bandwidth affects how fast megabytes move.
9. **Multimodal Dataset**
   Data combining text, images, audio, or video. Multimodal data uses large megabytes.
10. **Mutable Data**
    Data that changes during processing. Mutable data can increase temporary megabyte usage.

Letter N

1. **Network Bandwidth**
   The capacity of a network connection. Bandwidth controls how many megabytes can be transferred.
2. **Neural Network Size**
   The total storage required for a neural network. Size is measured in megabytes.
3. **Node Memory**
   Memory available on a computing node. Node memory limits megabytes per task.
4. **Normalization Data**
   Values used to scale data consistently. Normalization adds to dataset megabytes.
5. **Non-Volatile Memory**
   Memory that retains data without power. Capacity is measured in megabytes.
6. **Numerical Precision**
   The detail level of stored numbers. Higher precision increases megabytes.
7. **Noise Data**
   Unwanted or random information in datasets. Noise increases data size in megabytes.
8. **Neural Activation Size**
   Memory used to store neuron outputs. Activation size is measured in megabytes.
9. **Network Transfer Size**
   The amount of data sent across a network. Transfer size is tracked in megabytes.
10. **Node Cache**
    Temporary storage on a node for faster access. Cache size uses megabytes.

Letter O

1. **Object Storage**
   A storage method that manages data as objects. Capacity is measured in megabytes.
2. **Offline Dataset**
   Data stored locally rather than streamed. Offline datasets occupy megabytes.
3. **On-Device Storage**
   Storage located on the AI device itself. On-device capacity is measured in megabytes.
4. **Operational Memory**
   Memory used during active AI operations. Usage is tracked in megabytes.
5. **Optimization Data**
   Information used to improve model performance. Optimization data uses megabytes.
6. **Output Buffer**
   Temporary storage for results before delivery. Buffers hold megabytes.
7. **Overhead Data**
   Extra data required for system operation. Overhead increases total megabytes.
8. **Object Detection Output Size**
   The amount of data produced by detection models. Output size is measured in megabytes.
9. **Online Storage**
   Storage accessible via the internet. Online storage limits are measured in megabytes.
10. **Open Dataset Size**
    The storage footprint of publicly available datasets. Size is measured in megabytes.

Letter P

1. **Parameter Count**
   The total number of model values. Count correlates with megabytes required.
2. **Persistent Storage**
   Long-term storage that retains data. Capacity is measured in megabytes.
3. **Pipeline Buffer**
   Temporary storage between processing stages. Buffers manage megabytes in motion.
4. **Preprocessing Data**
   Data created while preparing datasets. Preprocessing adds megabytes.
5. **Precision Reduction**
   Lowering numeric detail to save space. Reduction decreases megabyte usage.
6. **Prediction Output Size**
   The amount of data produced by model predictions. Size is measured in megabytes.
7. **Primary Memory**
   Main system memory used for processing. Capacity is measured in megabytes.
8. **Processing Overhead**
   Extra resources needed during computation. Overhead increases megabyte usage.
9. **Profiling Data**
   Information collected to analyze performance. Profiling data uses megabytes.
10. **Paged Memory**
    Memory divided into fixed-size blocks. Paging manages megabytes efficiently.

Letter Q

1. **Quantization**
   A technique that reduces numerical precision to lower memory usage. Quantization helps AI models use fewer megabytes.
2. **Query Cache**
   Stored results from previous searches or requests. Query caches save megabytes by avoiding repeated computation.
3. **Query Latency**
   The time it takes to retrieve data after a request. Latency depends on how many megabytes must be accessed.
4. **Queue Buffer**
   A temporary holding area for data waiting to be processed. Buffers store megabytes in order.
5. **Quick Access Memory**
   Memory designed for fast retrieval. It handles megabytes needed immediately by AI tasks.
6. **Quantitative Dataset**
   Data made up of numerical values. These datasets are stored and measured in megabytes.

7. **Query Result Size**
   The amount of data returned from a query. Result size is measured in megabytes.
8. **Quorum Storage**
   A distributed storage method requiring agreement across systems. Data is replicated in megabytes.
9. **Quality Metrics Data**
   Information used to measure model performance. Metrics consume storage in megabytes.
10. **Quantized Model Size**
    The reduced size of an AI model after quantization. Size is typically measured in megabytes.

Letter R

1. **RAM Usage**
   The amount of active memory in use. RAM usage is tracked in megabytes.
2. **Read Throughput**
   The speed at which data is read from storage. Throughput affects how fast megabytes are accessed.
3. **Real-Time Processing**
   Processing data immediately as it arrives. Real-time systems manage megabytes continuously.
4. **Recursive Memory Use**
   Memory consumption during repeated function calls. Usage accumulates in megabytes.
5. **Redundant Storage**
   Extra copies of data kept for reliability. Redundancy increases total megabytes stored.
6. **Resource Allocation**
   Assigning system resources like memory and storage. Allocation is measured in megabytes.
7. **Runtime Memory**
   Memory required while a program is running. Runtime memory usage is tracked in megabytes.
8. **Raw Dataset Size**
   The size of unprocessed data. Raw datasets often require many megabytes.
9. **Read-Only Memory Footprint**
   The storage space used by non-modifiable data. Footprint is measured in megabytes.
10. **Retrieval Speed**
    How quickly stored data can be accessed. Speed depends on the number of megabytes involved.

Letter S

1. **Sample Size**
   The number of data examples used in AI training. Larger samples increase megabytes.

2. **Scalable Storage**
   Storage that grows with demand. Scalability supports increasing megabytes.
3. **Serialized Data Size**
   The storage size of data converted into a transferable format. Size is measured in megabytes.
4. **Shared Memory**
   Memory accessible by multiple processes. Shared memory usage is tracked in megabytes.
5. **Snapshot Size**
   The storage required for saved system states. Snapshots consume megabytes.
6. **Sparse Data Representation**
   A storage method that saves space when data contains many empty values. Sparsity reduces megabytes.
7. **Storage Capacity**
   The total amount of data a system can store. Capacity is measured in megabytes or more.
8. **Streaming Data Buffer**
   Temporary storage for incoming data streams. Buffers handle megabytes in motion.
9. **System Cache**
   Memory that stores frequently used data. Caches manage megabytes for faster access.
10. **Synchronization Data**
    Information used to keep systems aligned. Sync data adds to total megabytes.

Letter T

1. **Temporary Storage**
   Short-term storage used during processing. Temporary storage holds megabytes briefly.
2. **Tensor Size**
   The memory required to store multidimensional data structures. Tensor size is measured in megabytes.
3. **Throughput Rate**
   The amount of data processed per unit time. Rate is often described in megabytes per second.
4. **Training Dataset Size**
   The total amount of data used to train AI models. Size is measured in megabytes.
5. **Transfer Buffer**
   Temporary memory used during data movement. Buffers hold megabytes during transfer.
6. **Transactional Storage**
   Storage that ensures data consistency. Transactions add overhead measured in megabytes.
7. **Thread Memory Usage**
   Memory consumed by execution threads. Usage is tracked in megabytes.
8. **Token Storage Size**
   The memory required to store processed tokens. Token data uses megabytes.
9. **Time-Series Data Size**
   The storage footprint of data collected over time. Size increases in megabytes.
10. **Training Checkpoint Size**
    The storage required for saved training states. Checkpoints are measured in megabytes.

Letter U

1. **Unified Memory**
   A memory system shared between CPU and GPU. Unified memory manages data in megabytes efficiently across processors.
2. **Upload Size**
   The amount of data sent to storage or the cloud. Upload size is measured in megabytes.
3. **User Data Storage**
   Storage allocated for user-generated files. Capacity is tracked in megabytes.
4. **Utilization Rate**
   The percentage of memory or storage currently in use. Utilization reflects how many megabytes are occupied.
5. **Uncompressed Data Size**
   The size of data before compression. Uncompressed data often uses more megabytes.
6. **Update Payload**
   Data sent during software or model updates. Payload size is measured in megabytes.
7. **Uptime Storage Logs**
   Logs recorded while systems run continuously. Logs accumulate megabytes over time.
8. **Unified Dataset**
   A dataset combined from multiple sources. Size is measured in megabytes.
9. **User Session Memory**
   Memory allocated for active users. Session memory uses megabytes dynamically.
10. **Usage Quota**
    A storage limit set for users or systems. Quotas are defined in megabytes or larger units.

Letter V

1. **Validation Dataset Size**
   The amount of data used to evaluate AI models. Size is measured in megabytes.
2. **Vector Embedding Size**
   The memory required to store numerical representations of data. Size is measured in megabytes.
3. **Virtual Memory**
   A system that extends RAM using disk storage. Virtual memory is managed in megabytes.
4. **Volatile Memory**
   Memory that loses data when power is off. Capacity is measured in megabytes.
5. **Video Dataset Size**
   The storage required for video data. Video datasets consume many megabytes.
6. **Versioned Storage**
   Storage that keeps multiple file versions. Versioning increases megabyte usage.

7. **Vision Model Size**
   The memory footprint of computer vision models. Size is measured in megabytes.
8. **Vector Database Storage**
   Storage used for similarity search data. Vector databases consume large megabytes.
9. **Virtualized Storage**
   Storage abstracted from physical hardware. Capacity is tracked in megabytes.
10. **Volume Capacity**
   The total size of a storage volume. Capacity is measured in megabytes or more.

Letter W

1. **Weight Parameters**
   Learned values in AI models. Weight storage contributes to megabytes used.
2. **Working Memory**
   Memory actively used during processing. Working memory usage is measured in megabytes.
3. **Write Throughput**
   The speed at which data is written to storage. Throughput affects megabytes per second.
4. **Warm Cache**
   Cached data that is already loaded. Warm caches reduce repeated megabyte transfers.
5. **Workflow Data Size**
   The total data used across an AI workflow. Size is measured in megabytes.
6. **Web Dataset Size**
   The storage required for web-based data. Size is tracked in megabytes.
7. **Weight Compression**
   Reducing model weight size to save memory. Compression lowers megabytes used.
8. **Window Size (AI Context)**
   The amount of data processed at once. Window size affects memory megabytes.
9. **Writable Storage**
   Storage that allows data modification. Capacity is measured in megabytes.
10. **Workload Memory Demand**
   The memory required for tasks. Demand is expressed in megabytes.

Letter X

1. **X-Axis Data Size**
   Data stored along one dimension of a dataset. Size contributes to megabytes used.
2. **XML Data Size**
   The storage footprint of XML-formatted data. XML files often use many megabytes.
3. **XOR Encoding**
   A binary operation used in data processing. Encoded data occupies megabytes.

4. **Execution Context Size**
   Memory required during task execution. Size is measured in megabytes.
5. **External Dataset Size**
   The size of data sourced externally. External datasets add megabytes.
6. **Expanded Memory Use**
   Memory growth during processing. Expansion increases megabytes consumed.
7. **Experimental Dataset**
   Data used for testing AI models. Size is measured in megabytes.
8. **Extraction Buffer**
   Temporary storage during data extraction. Buffers store megabytes briefly.
9. **Extended Storage Capacity**
   Additional storage added to a system. Capacity is measured in megabytes.
10. **Explainability Data Size**
    Data used to explain AI decisions. Explainability outputs consume megabytes.

Letter Y

1. **Yield Data Size**
   The amount of useful output data produced. Yield is measured in megabytes.
2. **Yearly Data Growth**
   Annual increase in stored data. Growth is tracked in megabytes.
3. **YAML File Size**
   The storage footprint of YAML configuration files. Size is measured in megabytes.
4. **Y-Axis Feature Size**
   Feature data along a dataset dimension. Size contributes to megabytes.
5. **Young Generation Memory**
   Memory area for newly created data. Usage is measured in megabytes.
6. **Yield Optimization Data**
   Data used to improve output efficiency. Stored in megabytes.
7. **YourAIStudyBuddy Dataset Size**
   The total data used by AI study tools. Size is measured in megabytes.
8. **Yottabyte Reference Scale**
   A conceptual upper scale of data measurement. Used for comparison beyond megabytes.
9. **Yield Metrics Storage**
   Storage used for performance metrics. Metrics consume megabytes.
10. **Yoked Memory Allocation**
    Linked memory regions used together. Allocation is measured in megabytes.

Letter Z

1. **Zero-Copy Memory**
   A method that avoids duplicating data. Zero-copy reduces megabytes in motion.

2. **Zipped Dataset Size**
   The storage size after compression. Zipped files use fewer megabytes.
3. **Zettabyte Scale (Reference)**
   A large-scale data concept. Used to compare growth beyond megabytes.
4. **Zone-Based Storage**
   Storage divided into regions. Zones manage megabytes efficiently.
5. **Z-Score Normalization Data**
   Statistical values used in preprocessing. Data adds megabytes.
6. **Zero Padding**
   Adding placeholder data to inputs. Padding increases memory megabytes.
7. **Zoomed Feature Maps**
   Scaled representations in vision models. Stored in megabytes.
8. **Zonal Cache**
   Cache divided by region. Cache size is measured in megabytes.
9. **Zero-Day Model Update Size**
   The storage required for emergency updates. Size is measured in megabytes.
10. **Zipped Model Archive**
    Compressed AI model files. Archives reduce megabytes required.

Student Benefits

Why This Book Matters

Artificial intelligence is transforming the world at an unprecedented pace. From healthcare and transportation to education and scientific research, AI is becoming an essential part of everyday life. Understanding how intelligent systems work is no longer a specialized skill—it is becoming an important component of digital literacy for students of all ages.

The Future of Intelligence: Humans and Artificial Minds provides students with a strong foundation for understanding the technologies that are shaping tomorrow's society. Rather than focusing only on technical definitions, this book encourages students to think critically about how humans and intelligent machines can work together responsibly and ethically.

Throughout the sixteen chapters, students will explore the evolution of intelligence, the relationship between human and artificial thinking, emerging AI technologies, and the opportunities and challenges that accompany rapid innovation. Real-world examples, chapter summaries, reflection activities, worksheets, quizzes, and answer keys reinforce learning while encouraging independent thinking.

By completing this book, students will be able to:

- Develop a foundational understanding of artificial intelligence and future intelligent systems.

- Explain the similarities and differences between human intelligence and machine intelligence.
- Build AI vocabulary and strengthen digital literacy skills.
- Analyze ethical, social, educational, and economic impacts of AI.
- Improve critical thinking and problem-solving abilities.
- Recognize the importance of responsible AI development and use.
- Prepare for future careers where AI will play an increasingly important role.
- Build confidence in discussing emerging technologies using accurate terminology.
- Connect AI concepts to real-world applications across multiple industries.
- Continue learning as AI technologies evolve throughout their education and careers.

As students progress through each chapter, they will discover that understanding intelligence is not simply about learning how machines work. It is also about understanding human creativity, decision-making, collaboration, and the important role people continue to play in shaping the future of technology.

Learning AI today helps prepare students to become informed citizens, responsible technology users, creative problem-solvers, and future innovators.

Teacher Implementation Guide

Using This Book Effectively

This book is designed to be flexible and can be implemented in middle school, high school, career and technical education (CTE), introductory college courses, adult education, or independent study.

Suggested Instructional Sequence

Each chapter follows a consistent learning cycle:

1. Read the instructional chapter.
2. Review the chapter summary.
3. Complete the reflection activity.
4. Finish the worksheet.
5. Take the chapter quiz.
6. Review the answer key and discuss any misconceptions.

Estimated Instructional Time

- Chapter Reading: 30–45 minutes
- Classroom Discussion: 15–20 minutes
- Reflection Activity: 10–15 minutes

- Worksheet: 20–30 minutes
- Quiz: 10–15 minutes
- Review and Feedback: 10–15 minutes

Each chapter may be completed during one extended instructional block or divided across multiple class sessions.

Differentiated Instruction

Teachers may adapt lessons by:

- Assigning selected chapters for introductory learners.
- Using group discussions for collaborative learning.
- Providing additional examples for students requiring support.
- Encouraging independent research for advanced learners.
- Connecting AI concepts to current events and emerging technologies.

Classroom Discussion Topics

Throughout the course, encourage students to explore questions such as:

- How does AI compare with human intelligence?
- What responsibilities accompany powerful AI technologies?
- How can AI improve society?
- What ethical challenges should future developers consider?
- Which careers will require AI literacy?

Course Outcome

Upon completion of all sixteen chapters, students should possess a foundational understanding of artificial intelligence, future intelligent systems, ethical considerations, and the growing role AI will play across education, business, healthcare, science, and society.

This instructional design provides teachers with a complete, organized framework for introducing AI concepts while promoting critical thinking, responsible technology use, and lifelong learning.

Teacher Implementation Guide

Using This Book Effectively

This book is designed to be flexible and can be implemented in middle school, high school, career and technical education (CTE), introductory college courses, adult education, or independent study.

Suggested Instructional Sequence

Each chapter follows a consistent learning cycle:

1. Read the instructional chapter.
2. Review the chapter summary.
3. Complete the reflection activity.
4. Finish the worksheet.
5. Take the chapter quiz.
6. Review the answer key and discuss any misconceptions.

Estimated Instructional Time

- Chapter Reading: 30–45 minutes
- Classroom Discussion: 15–20 minutes
- Reflection Activity: 10–15 minutes
- Worksheet: 20–30 minutes
- Quiz: 10–15 minutes
- Review and Feedback: 10–15 minutes

Each chapter may be completed during one extended instructional block or divided across multiple class sessions.

Differentiated Instruction

Teachers may adapt lessons by:

- Assigning selected chapters for introductory learners.
- Using group discussions for collaborative learning.
- Providing additional examples for students requiring support.
- Encouraging independent research for advanced learners.
- Connecting AI concepts to current events and emerging technologies.

Classroom Discussion Topics

Throughout the course, encourage students to explore questions such as:

- How does AI compare with human intelligence?
- What responsibilities accompany powerful AI technologies?
- How can AI improve society?
- What ethical challenges should future developers consider?
- Which careers will require AI literacy?

Course Outcome

Upon completion of all sixteen chapters, students should possess a foundational understanding of artificial intelligence, future intelligent systems, ethical considerations, and the growing role AI will play across education, business, healthcare, science, and society.

This instructional design provides teachers with a complete, organized framework for introducing AI concepts while promoting critical thinking, responsible technology use, and lifelong learning.

NOTES:

NOTES:

NOTES:

www.ingramcontent.com/pod-product-compliance
Lightning Source LLC
Chambersburg PA
CBHW041754220926
51847CB00014B/198
* 9 7 8 1 9 7 2 9 2 5 1 0 2 *